Role of Cell Surface in Development

Volume I

Author

K. Vasudeva Rao, Ph.D.
Professor
Department of Zoology
University of Delhi
Delhi, India

CRC Press, Inc.
Boca Raton, Florida

Library of Congress Cataloging-in-Publication Data

Vasudeva Rao, K., 1933—
  Role of cell surface in development.

  Bibliography: p.
  Includes index.
  1. Cell membranes. 2. Developmental cytology.
I. Title.
QH601.V385   1987   574.3   86-13684
ISBN-0-8493-4687-8 (set)
ISBN-0-8493-4688-6 (v. 1)
ISBN-0-8493-4689-4 (v. 2)

Direct all inquiries to CRC Press, Inc., 2000 Corporate Blvd., N.W., Boca Raton, Florida, 33431.

© 1987 by CRC Press, Inc.

International Standard Book Number 0-8493-4687-8 (set)
International Standard Book Number 0-8493-4688-6 (Volume I)
International Standard Book Number 0-8493-4689-4 (Volume II)

PREFACE

Developmental biology has now emerged as a truly interdisciplinary science. Classical embryology of past centuries has revealed a wealth of knowledge on how animal embryos develop. From this knowledge comparative embryologists have reached such far-reaching generalizations as the homology of germ layers. Research in embryology during the present century is characterized by the experimental approach. Interfering with developing embryos in various ways has been the essential feature of the work of experimental embryologists. In the past two decades many new techniques developed by biochemists and biophysicists have contributed greatly to understanding the mechanisms of development. A developmental biologist can now expect to find a rational explanation for embryological observations which once looked mysterious. Progress in a multidisciplinary science needs to be surveyed periodically, not only to take stock of the achievements, but also to indicate profitable future approaches. A restatement of old problems often brings into sharper focus the area where rapid progress is possible with the newer techniques.

Recent progress in the area of membrane biology amply justifies a discussion of animal development to highlight the role of the cell surface. I have had to survey a fairly wide area of modern biology in order to highlight the role of the cell surface in development. Consequently I had to be selective in the choice of suitable examples of developing systems for the discussion. This might have left many lacunae in the citation of contributions from various workers. In general the bibliographic references are intended to be informative to the reader rather than to give credit to individual scientists who published the original reports. Thus not all publications on a given finding are cited. I apologize to the scientists whose important work, in spite of being relevant, is not mentioned in my narration.

I owe a great deal of thanks to several persons for different reasons. My wife, Sunanda, has been an unending source of inspiration in all my efforts. I thank her for bearing with the various problems which arose during the preparation of this book. I thank my sons, Vinay and Ashok, who helped me with enthusiasm in comparing the final typescript with the original draft. Dr. Shashanka Bhide deserves special thanks for his help in comparing the final typescript. I take this opportunity to thank my teacher, Professor Leela Mulherkar, who introduced me to this fascinating subject. In the preparation of the illustrative material, I received help from Mr. R. K. Bhandari and Mr. E. A. Daniels. Mr. K. V. L. N. Rao, the administrative officer of the department, has rendered help in various ways. Mr. K. R. Sharma, the librarian of the department, deserves a special mention for help in locating a number of references. He often pursued the matter even after I had given up. Mr. Baldev Singh Rana has exhibited remarkable patience in typing the manuscript. I am thankful to all these persons.

I am greatly indebted to Dr. G. V. Sherbet and Prof. Ruth Bellairs who read some parts of the manuscript. Their criticism and suggestions have contributed to improving the text considerably. In spite of the help obtained from them, however, I own complete responsibility for any errors and omissions in the book. It was a pleasure to work in association with Mr. B. J. Starkoff of CRC Press and his colleagues during the production of the book. I am thankful to the authors and publishers of various books/journals from whom I have borrowed material. These are acknowledged individually wherever such contributions have been used.

My research efforts in the past two decades have been aided financially by various agencies including the Council of Scientific and Industrial Research, The Indian Council of Medical Research, the Indian National Science Academy, and the Department of Science and Technology, Government of India.

Delhi
September 1985

K. Vasudeva Rao

THE AUTHOR

Dr. K. Vasudeva Rao is and has been a teacher and researcher in the University of Delhi where he has given developmental biology courses at the undergraduate and Master's level for more than twenty years. He is especially interested in testing and evaluating material for introducing new exercises in developmental biology for undergraduate and Master's courses.

Dr. Rao started his research as a student of the eminent Indian embryologist Prof. Leela Mulherkar at the University of Poona. After obtaining his Ph.D. degree, he moved to the University of Delhi as a lecturer. Since 1983 he has been a Professor. He has published more than forty papers and three reviews. His current research programs are intended to elucidate the cellular basis of morphogenetic changes. Dr. Rao's research has received support from the Indian National Science Academy, New Delhi, and the central government funding agencies such as the Council of Scientific and Industrial Research, the Indian Council of Medical Research, and the Department of Science and Technology (Government of India).

Dr. Rao is a member of the Indian Science Congress Association, the Indian Society of Cell Biology, and the Indian Society of Developmental Biologists.

With grateful thanks and respect to my teacher

PROFESSOR LEELA MULHERKAR

TABLE OF CONTENTS

Volume I

Chapter 1

THE PLASMA MEMBRANE: ITS STRUCTURE AND ATTRIBUTES

I. INTRODUCTION

The development of embryos is one of the most fascinating biological phenomena. The process begins with a rapid division of the zygote, giving rise to a large number of cells derived from it. Following this is a large scale rearrangement of the cells. They migrate from one region to another, moving over long distances, and following amazingly precise courses. The cells resulting from the cleavage are far from identical. Eventually they diversify further, assume different shapes, elaborate different substances, and organize themselves into tissues and organs of the embryo. Descriptive embryology has revealed these fascinating changes in great detail. Experimental interference with embryonic development has given us some more insight into the process. Experimental embryologists have shown that cells interact continuously among themselves during development. These interactions lead to extensive reprograming of the synthetic activities in the cells. Understanding the mechanisms of developmental change has been facilitated greatly by the progress in other related fields such as biochemistry and cell biology. An analysis of the highly complex process of embryogenesis necessarily takes recourse to the techniques and concepts developed in the related biological and physical sciences. Depending on one's interest (and competence), any one of several possible approaches may be followed in analyzing the mechanisms of embryogenesis. Eventually, the information obtained from the different approaches is synthesized to obtain a comprehensive picture. Surgical or chemical interference with embryos results in abnormal development, suggesting the mechanisms underlying normal development. Differentiation of the cells is essentially a consequence of precisely programed differential gene expression. Impressive progress has been made in discovering the molecular events occurring within the cells and leading to the elaboration of some specific gene products characterizing the differentiated cell phenotype. Though the biochemical changes leading the cells into diverse pathways of differentiation occur within the cells themselves, some external signals are necessary to initiate these changes. The signals are received at the cell periphery. Classical experimental embryology has shown that the external signals in most cases are of the nature of mutual interactions of cells that are in the close vicinity of each other. The interactions, commonly called embryonic inductions, are as diverse as numerous. The nature of the signals exchanged by the cells has been hotly pursued for several decades. However, not much has emerged from these efforts.

Inasmuch as the inductive signals are exchanged at the cell surface, a detailed knowledge of the surface structure and function is imperative. The consequences of differentiation include not only the elaboration of specialized products within the cytoplasm but also the acquisition of new properties by the cell surface. It is now abundantly clear that the cell surface is as important as the nucleus in regulating the activities of the cells. A study of the role of the cell surface in developmental processes is therefore a specially profitable avenue of approach to understanding embryogenesis. This approach was clearly defined by Curtis[1] nearly two decades ago.

Recent progress in cell biology has yielded a wealth of information that can aid in understanding embryonic development in a clearer perspective. The specified objective of this book is to show that the cell surface plays an important role in the development of animal embryos. We shall have to survey a fairly wide field of biological research that aims at revealing the mechanisms underlying cellular activities. As an essential background, the molecular architecture of the plasma membrane will be described first. In subsequent chapters we shall discuss some generalized properties of living cells which have a direct bearing on the problem to be discussed. These properties include cell adhesiveness and motility. Fusion of cells is a highly specialized developmental process in which the cell surface is directly involved. Cell fusion is met with in fertilization, the formation of myotubes, and some other, less spectacular, situations. We shall then select a few simple examples exhibiting developmental changes: the cellular slime molds and aggregates of sponge cells or the cells obtained from embryonic organs. The cellular activities in them are akin to some developmental processes met within embryos. They are therefore suitable for experimental manipulation which can reveal some mechanisms of development. This will be followed by some specific examples of embryonic developmental processes to highlight the role of the cell surface. Finally, a brief discussion will be presented on the concept that cancer is a disease related to the developmental processes and that the cell surface plays a definite role in it.

II. THE PLASMA MEMBRANE

The outermost limiting membrane of the cell is variously known as the plasma membrane, cell surface membrane, or simply the cell surface. Calling it a surface can imply just a two-dimensional area of the cell without including its thickness. However, the term *cell surface* has been used synonymously with *plasma membrane* for a long time, and it is now difficult to stop its usage. The term *cell membrane,* on the other hand, implies the entire membrane system of the cell (the intracellular as well as the cell surface). It is therefore advisable to avoid the use of the term *cell membrane* when one would like to refer particularly to the cell surface membrane. Whatever the terminology one likes to use, the meaning has to be clarified. By the terms plasma membrane, cell surface membrane, and cell surface, we mean the limiting membrane of the cell, which consists of a lipid bilayer with its integral components. In this chapter, we shall deal with the molecular architecture of the plasma membrane and the biophysical properties resulting therefrom. That the plasma membrane should have a role in regulating the entry of substances into and their exit from the cell is obvious. An elucidation of this role of the plasma membrane would depend on a clear understanding of its structure and chemical composition. Theories proposed on the molecular organization of the plasma membrane have to account for the known properties of this cell organelle. From time to time, different theories have been proposed in this connection, and we shall examine briefly the important ones. Our knowledge has been enriched by investigations that examined the various molecular models of the plasma membrane and added to them certain modifications to account for its known properties.

An understanding of the molecular structure of the plasma membrane may be considered as having started from the turn of the last century when Overton[2] showed that the entry of many molecules into the cell depended on their lipid solubility. If the plasma membrane is rich in lipids, this property can be explained easily. This suggestion has proved to be cardinal for all the theories proposed subsequently on the structure of the plasma membrane. Attempts were then made to obtain lipids form the plasma membrane. In the early studies, the erythrocyte plasma membrane was used in view of the ease in obtaining and handling it. Mammalian erythrocytes are nonnucleated cells, and when lysed in hypotonic media, they yield large pieces of the plasma membrane without significant contamination of the cytoplasmic components. The membranes thus obtained yield lipids on acetone extraction, vindicating the suggestion of Overton. Görter and Grendel[3] obtained lipids from the erythrocytes of several mammalian species and estimated them quantitatively. The area covered by a single molecular layer of the lipids was found to be twice the area of the cell surface, suggesting that the plasma membrane is a double lipid layer. Though the estimates were only approximate and included sources of error as pointed out subsequently,[4] the conclusion

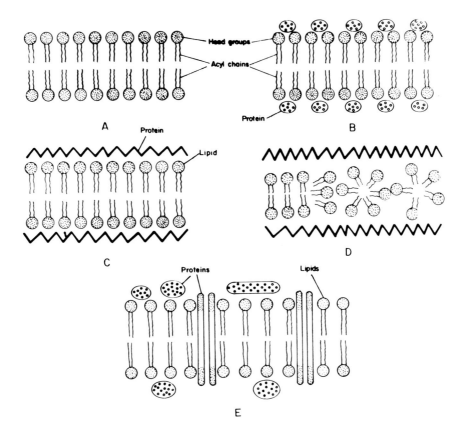

FIGURE 1. "Classical" molecular models proposed for the structure of plasma membrane. (A) a lipid bilayer; (B to D) proteins on either side of the lipid bilayer showing different arrangements of the protein and lipid molecules. (E) a membrane model with proteins peripheral to the lipid bilayer as well as spanning it. See text.

drawn from the study is still valid. Most of the subsequent models of the plasma membrane structure have been based on this assumption. Danielli and Davson[5] elaborated the double lipid layer model by adding a protein component on its outer and cytoplasmic sides. This was to account for the finding that the observed surface tension of the plasma membrane was lower than considered possible for a lipid layer. To account for the active transport of ions and passive diffusion of water soluble substances, protein coated channels spanning the membrane have been postulated. These "classical" models of the plasma membrane structure (Figure 1) were largely based on the information obtained from the analysis of erythrocyte membrane preparations and were intended to account for the permeability properties of living cells. Current ideas on the structure of the plasma membrane take into account the physicochemical properties of the constituent molecules and a variety of experimental observations on living cells. We shall first describe the chemical constituents of the plasma membrane and their elaboration by the cell. It is hoped that this will provide us with the necessary background to comprehend the molecular organization of the plasma membrane in the right perspective.

A. Chemical Composition of the Plasma Membrane

The first step in revealing the molecular architecture of the plasma membrane is to separate and characterize all the chemical components constituting it. Biochemical skill of the highest order has been applied to achieve this goal. Highly sophisticated instrumentation and the development of efficient separation methods have enabled a detailed analysis of the various

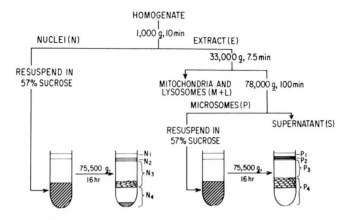

FIGURE 2.   An experimental schedule for the preparation of plasma membranes of hepatocytes. (From Touster, O., Aronson, N. N., Jr., Dulaney, J. T., and Hendrickson, H. Reproduced from *The Journal of Cell Biology,* 1970, 47, 604 by copyright permission of The Rockefeller University Press.)

constituents. The physical and chemical properties of the component molecules give clues to the nature of their association and organization for functional efficiency. The synthesis of the membrane components is another aspect of the study. Recent work in this field has provided a fairly accurate description of the molecular organization of the plasma membrane.

1. Cell Fractionation and Isolation of Plasma Membranes

One of the serious shortcomings of the work using erythrocytes is the fact that all the information on the chemical constituents of the plasma membrane pertains to an atypical cell. Methods were therefore evolved to obtain membrane preparations from a variety of other tissue cells. It should be emphasized that no such thing as a standard method for plasma membrane preparation exists. Each species, and even the developmental or pathological state of a given tissue, has unique characteristics that necessitate developing special procedures of preparation. Regardless of the sophistications introduced, most of these methods yield "membrane enriched" fractions from disrupted cells. As an example, the recommended schedule for isolating rat hepatocyte membranes is presented diagrammatically in Figure 2.

The major steps involved in the preparative isolation of plasma membranes are cell rupture and differential centrifugation. It is needless to emphasize that the procedures should be mild and least traumatic to the cell organelles. Homogenization in 0.25 M sucrose has been a widely used method. Using hypertonic sucrose (0.88 M) or hypotonic media has also been recommended in some cases. The composition of media, the rotors, and speed of centrifugation are precisely defined. Successful isolation largely depends on strictly following the procedures laid down by those who developed the method. The theoretical aspects of the techniques used are beyond the scope of this text. Methods applicable to a wide variety of tissue cells have been compiled by Fleischer and Packer,[7] and many more methods have been published subsequently. As a matter of fact, so many different methods for plasma membrane preparation have been described that it is often difficult to choose a suitable one for a given tissue. Often new methods are described without any tangible advantage over existing ones. The frustration generated by such a situation is indicated by the suggestion from a well-known worker in the field that a moratorium be imposed on the publication of "new" methods unless there is *real* improvement over the existing ones.[8] In general, methods applicable to solid tissues are not suitable for cells in suspension or in monolayers in vitro. For a detailed discussion on the physical principles involved in cell fractionation, reference may be made to Beaufy and Amar-Costesec.[9]

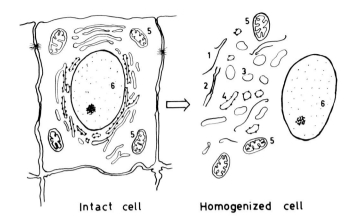

Intact cell   Homogenized cell

FIGURE 3.   Various organelles obtained when a cell is homogenized. 1, A sheet of plasma membrane; 2, membrane sheet with a junctional complex; 3, membrane vesicle; 4, microsome; 5, mitochondrion; 6, nucleus.

Disruption of cells results in the release of plasma membrane fragments of various sizes. Mechanical shearing of cells using homogenizers is the most commonly used method to achieve cell rupture. A highly satisfactory and "gentle" procedure of cell disruption is what is called "nitrogen cavitation" method. In this technique, cell suspensions are first equilibrated with 30 to 80 atm of nitrogen gas and then returned to atmospheric pressure with minimum liquid shear. For a review on cell disruption techniques see Wallach and Winzler.[10]

The membrane system of cells also includes the intracellular membranes (endoplasmic reticulum, Golgi complex, etc.), and these too are obtained in the membrane preparations from disrupted cells. Obtaining the plasma membrane excluding the other membrane structures offers many difficulties. Most of the procedures developed to overcome these difficulties are chiefly empirical. When cells are ruptured, the cytoplasmic membranes have a tendency to form small vesicles called microsomes. Small fragments of the plasma membrane also may close into vesicles (Figure 3), and during the subsequent procedures of centrifugation, they may be found with the microsomes. Certain features such as cell junctions have the effect of strengthening the plasma membrane and consequently yielding large sheets of it instead of vesicles. Controlling the shear applied during cell disruption can also achieve the same result. Some workers have used -SH (sulfhydryl) blocking agents for strengthening the plasma membrane.[11] In the absence of special strengthening structures such as junctional complexes or experimentally induced stability, vesicles of plasma membranes are found along with cytoplamic membranes. This offers special difficulties since they tend to remain together in centrifugation fractions. Strengthening the plasma membrane, albeit involving a risk of altering some membrane components (such as enzymes), has therefore been considered necessary for obtaining satisfactory preparations for biochemical analysis.

Enrichment of the plasma membrane fraction during preparation by centrifugation or other methods is generally monitored by the enrichment of a membrane associated substance such as an enzyme (Table 1). One of the most frequently used plasma membrane marker enzymes is 5'-nucleotidase (EC 3.1.3.5). This may not be suitable for some particular tissues, and other enzymes have to be chosen. For diverse lymphocytes, alkaline *p*-nitrophenyl phosphatase (EC 3.1.3.1) and Mg^{2+}-stimulated Na^+, K^+-dependent ATPase (EC 3.6.1.3) have been used.[12] Microsome vesicles may be monitored by NADH-oxidoreductase (EC 1.6.4.3) and glucose-6-phosphatase (EC 3.1.3.9). Possible contamination of the preparations by mitochondria may be checked by the assay of the mitochondrial enzymes, succinate dehydrogenase (EC 1.3.99.1), and monoamine oxidase (EC 1.4.3.4). Lysosome contamination of the plasma membrane fraction may be monitored by β-glucuronidase (EC 3.2.1.31).

Table 1
CHANGES IN THE ENZYME CONTENT OF DIFFERENT
CENTRIFUGAL FRACTIONS OF DISRUPTED RAT HEPATOCYTES

	Fraction[b]			
Enzyme[a]	N	M + L	P	S
5′-Nucleotidase	22.8 ± 4.0	12.9 ± 2.6	54.5 ± 2.1	10.9 ± 4.5
Phosphodiesterase I	17.4 ± 3.4	13.5 ± 4.8	50.9 ± 60	4.5 ± 1.6
Glucose-6-phosphatase	7.0 ± 2.6	22.0 ± 2.2	68.1 ± 2.5	5.3 ± 1.9
Cytochrome oxidase I	7.2	79.8	1.2	0
Succninate-cytochrome C reductase	6.6 ± 0.9	96.0 ± 4.7	5.3 ± 5.6	0

[a]   Enzyme quantities are expressed as percent of total in the homogenate. For details see Touster et al.[6]
[b]   Fractions as shown in Figure 1.

Morphological criteria such as brush borders of intestinal or kidney tubule epithelial cells can also be used where appropriate. Electron microscopic examination is often resorted to as the characteristic ''staining'' and dimensions of the plasma membrane permit visualization of the enriched fraction. Such monitoring is particularly valuable where regional differences in the membranes of cells are obvious, as in the luminal surface of intestinal epithelial cells with the brush border. It is indeed known that the chemical composition of the membrane differs between the brush border and the rest of the membrane.[13]

Having obtained a satisfactory ''membrane enriched'' fraction, the next step is to extract the different molecular constituents. Whereas lipids do not offer any special problems in this connection, the proteins are found to be very difficult to handle. This is so because they (the proteins) are relatively larger molecules, and the extent of their association with the lipids also varies widely. Proteins, which are only loosely associated with the plasma membrane but nevertheless important for its function (and hence justifiably considered as membrane components), are likely to be lost or only partly retained during the process of membrane preparation. On the other hand, proteins that are an integral part of the membrane could be modified as a result of their dissociation from the interaction with other components of the membrane. The domains of the protein molecules that are associated with the membrane lipids might change when exposed to an aqueous environment.

The possible rearrangements in the structure of membrane proteins have been discussed by Maddy and Dunn[14] and are depicted in Figure 4. Lytic enzymes released during cell rupture may degrade the proteins. The hydrophobic domains may come together due to repulsion from the aqueous milieu and thus several like or unlike protein molecules may form aggregates. More subtle conformational changes are also likely to occur after dissocation from the lipids. Labile molecules such as enzymes may show decreased activity, presumably because of their dissociation from the lipid environment. For example, the activity of sodium/potassium dependent ATPase of the red blood cell membrane is restored by the replacement of lipid.[15] There are also cases of increased enzyme activity due to preparative procedures. The involved techniques developed for cell fractionation and analysis of membrane preparations attempt to guard against the various pitfalls mentioned above.

Membrane proteins can be extracted from the enriched preparations using various methods for their solubilization. These include (1) modification of the electrostatic conditions in the membrane's environment (low ionic strength with or without chelating or reducing agents, high ionic strength, manipulation of pH); (2) modification of ionogenic groups within the membranes; (3) addition of protein perturbants; (4) treatment with enzymes; (5) sonication; (6) extraction with organic solvents, and (7) use of nonionic detergents. In addition, there are special methods applicable to glycoproteins. A detailed discussion of these methods and

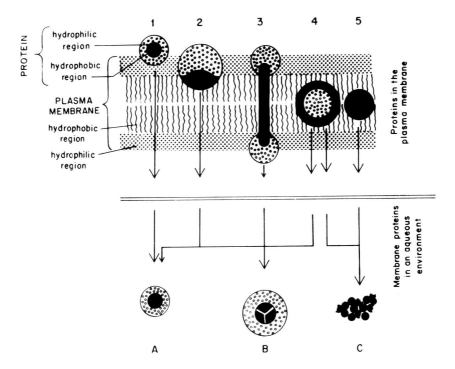

FIGURE 4.   Possible arrangements of proteins relative to a lipid bilayer and the probable effects of extraction of the proteins into aqueous solutions. 1, Proteins and lipids interact exclusively through their hydrophilic regions and the protein can be transferred into an aqueous environment with the minimum perturbation (state A). 2, Proteins and lipids interact through both hydrophilic and hydrophobic regions and on separation and exposure of the hydrophobic facets of the proteins to water they rearrange, either within a molecule (state A) or as a multimolecular complex (state B). 3, Here the protein traverses the lipid but the consequences of aqueous solubilization are as in 2 (i.e., states A or B). 4, Interaction between the protein and lipid is purely hydrophobic as the hydrophilic amino acids have been internalized (a possible but improbable arrangement). In an aqueous environment, the protein could invert to states A or B, or could form an insoluble conglomerate as in state C except that the subunits would have hydrophilic cores, or a mixed complex of B and C could arise. 5, The protein is composed exclusively of hydrophobic amino acids and would form an insoluble conglomerate in aqueous environments (state C). A protein of this type would be soluble in nonaqueous media. It is assumed in this scheme that proteins are incorporated into membranes by direct interaction with lipids. Some membrane proteins may not, in nature, interact with lipid, but be bound to the membrane by interaction with those proteins that do. This additional possibility extends, but does not alter, the concepts illustrated in the diagram. (From Maddy, A. H. and Dunn, M. J., *Biochemical Analysis of Membranes,* A. H. Maddy, Ed., Chapman & Hall, London, 1976, chap. 6. With permission.)

their applications is outside the scope of this text; however, such information can be obtained from Maddy and Dunn.[14]

An important problem in the isolation and characterization of the plasma membrane components is the contamination of the membrane enriched fractions by cytoplasmic or extracellular molecules. In an attempt to overcome this, various methods of labeling the membrane components have been developed. If a membrane component is suitably labelled *in situ* (before disrupting the cell), it can be followed in the purification procedure. The presence of such label on the molecules would indicate that they are indeed membrane components and not cytoplasmic contaminants. The choice of a suitable label depends on the type of tissue cells and the specificity with which it binds the membrane components. In general, such a label should bind permanently and covalently on the membrane component under physiological conditions, without penetrating into the cell. Many such chemical labels have been used. They label the amino, sulfhydryl, or other groups of the membrane molecules (see Table 2). Labeling of tyrosine residues of the exposed membrane proteins by radio-

Table 2
SOME EXAMPLES OF MEMBRANE LABELS

No.	Reagent	Reactive group in the membrane	Remarks
1	4-Acetamido-4′-iso-thiocyano-2-2′-stilbene disulfonate (SITS)	$-NH_2$	Impermanent; noncovalent
2	Para-chloro mercuribenzene sulfonate (PCMBS)	$-SH$	Slowly permeates; noncovalent
3	1-Fluoro-2,4-dinitro benzene (FDNB)	$-SH$, $-NH_2$ and tyrosine residues	Permeates; covalent
4	N-Ethyl maleimide (NEM)	$-SH$	Permeates; covalent
5	2-4-6-Trinitrobenzene sulfonate (TNBS)	$-NH_2$	Permeates slowly; covalent
6	Carboxypyridine disulfide	$-SH$	Does not permeate at low concentrations; covalent; introduces a $-COOH$ group

iodine in the presence of lactoperoxidase (EC 1.11.1.7) has been used extensively. The enzyme, being a macromolecule, does not diffuse into the cytoplasm of living cells, and thus the labeling is restricted to the exposed tyrosine residues of the membrane proteins (Figure 5). The cells are then disrupted, and the proteins obtained from the membrane enriched fractions are subjected to various methods of separation. Labeled components are detected by a suitable method (liquid scintillation counting or exposure to a sensitive film after separation on a gel). Many workers have employed this method for isolation and characterization of membrane components of a wide variety of cells. The technical aspects of this are somewhat intricate;[12,16] nonetheless, it may be considered as the method of choice in most situations, especially when one is looking for the appearance or disappearance of specific membrane proteins during the process of cell differentiation or any physiological change. Serine and threonine residues of membrane proteins can be labeled by $(\gamma^{32}P)$-ATP in the presence of phosphoprotein kinase. Terminal galactose residues in glycoproteins/glycolipids can be labeled using NaB^3H_4 in the presence of galactose oxidase[17,18] or by addition of a tritiated sialyl residue mediated by the sialyl transferase.[19] Some of these methods of labeling have been depicted in Figure 5.

Membrane lipids are generally extracted using a mixture of a relatively apolar lipid solvent such as chloroform and a polar alcohol. Chloroform-methanol mixtures are the most frequently used lipid extractants. Various techniques for the extraction and analysis of membrane lipids are reviewed by Veerkamp and Broekhuyse.[20] Carbohydrates are found as glycoproteins or glycolipids in the plasma membrane. They are thus obtained with the protein or lipid fractions. Methods for the isolation of glycoproteins and glycolipids and their analysis are reviewed by Cook.[21]

Chemical analysis of the plasma membrane has yielded adequate information leading to the conclusion that it is a mosaic of lipids and proteins, some of which bear oligosaccharide residues covalently bound to them. Lipids constitute 30 to 40% of the membrane components and proteins account for 50 to 60%. By weight, the carbohydrate component of the membrane is 1 to 8%. The relative amounts of these membrane constituents are variable according to the cell type. Protein to lipid to sugar ratios of human erythrocyte are 49:43:8, whereas those of myelin sheath are 18:79:3. Obviously the differences are related to the cells' functions.

a. Lipids

Membrane lipids are arranged as a double layer. There is good evidence to infer that the hydrophilic polar ends of the lipids are directed towards the cell surface and the cytoplasm. The hydrophobic nonpolar parts of the molecules must therefore be directed towards each

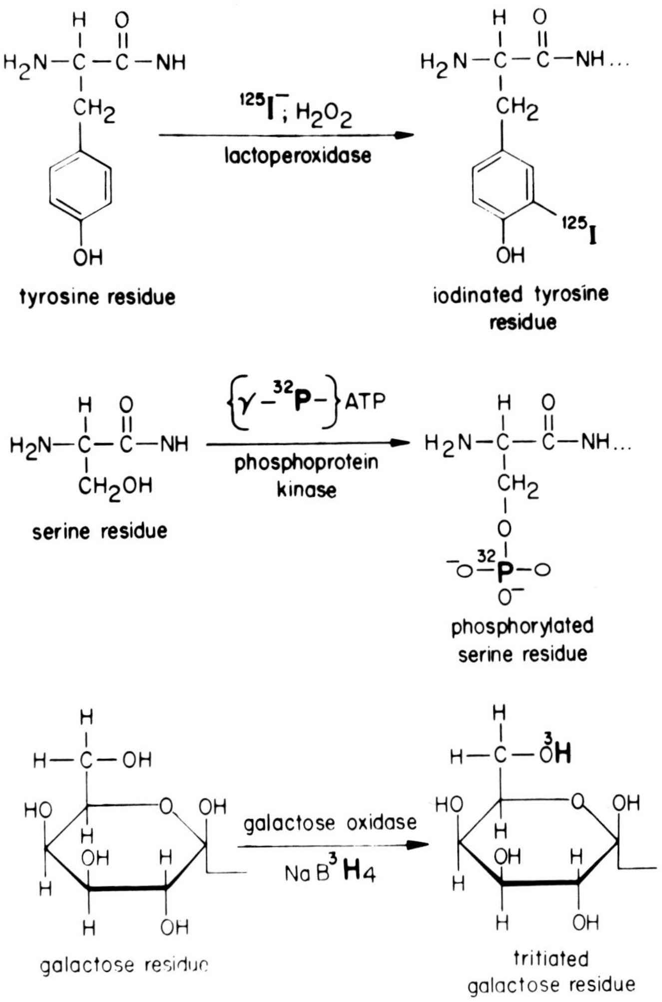

FIGURE 5. Different methods of radiolabeling cell surface molecules. The radioisotopes are shown in thick black letters. The labeling of the sugar, galactose, involves first, oxidation of the $-OH$ to $-CHO$ and then its reduction by NaB^3H_4 to incorporate the tritium atom.

other, forming a hydrophobic region in the plasma membrane. When lipids are spread on the water-air interface, the hydrophilic polar groups rest on the water surface while the hydrophobic part is repelled and stands projecting into the air. Evidence from physical and physicochemical studies[22-24] has shown that the lipid bilayer concept is correct.

Based on their chemical structure, the membrane lipids may be grouped as (1) phosphoglycerides, (2) sphingolipids, and (3) sterols. The building blocks of phosphoglycerides are glycerol, some other kinds of alcohols, and fatty acids. One of the hydroxyl groups of glycerol is esterified with phosphoric acid and the other two with long fatty acid chains. One of the other alcohols (ethanolamine, choline, serine or inositol) is esterified with the phosphoric acid (Figure 6). The negatively charged phsophate group and the other ionizable groups on the alcohols, if present, render this end of the molecule hydrophilic in nature; the fatty acyl chains are, on the contrary, hydrophobic. Thus the phosphoglycerides are, as a whole, amphipathic. Sphingolipids are similar to the phospholipids in having two hydrophobic tails and a polar group. However, they have no glycerol. Sphingolipids are of three

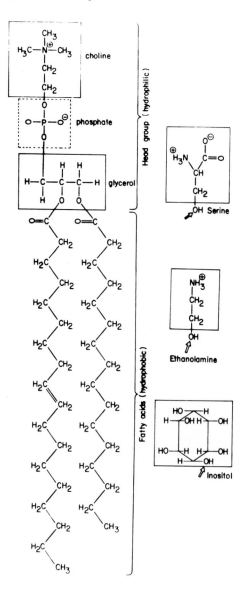

FIGURE 6.   The structure of phospholipids. The molecular formula shown on the left is phosphatidyl choline with the fatty acid chains, oleic acid (left) and palmitic acid (right) esterified to the glycerol. The different parts of the structural formula are boxed for clarity. Choline may be replaced by the formulas shown on the right to depict phosphatidyl-serine, ethanolamine, or inositol. The hydroxyl that esterifies with the phosphate is shown by an arrow.

subclasses: sphingomyelins, cerebrosides, and gangliosides. Sphingomyelin, a phospholipid, consists of sphingosine with one of its -OH groups esterified with phosphorylated choline and the other two groups with hydrophobic units (Figure 7). Cerebrosides contain no charged head groups; they characteristically consist of one or more monosaccharide units (Figure 8). Gangliosides are the most complex sphingolipids, containing very large polar groups made up of several sugar units. Usually one or more of the terminal sugar units is N-acetylneuraminic acid (sialic acid), which has an ionizable -COOH group (Figure 8). Cholesterol is the major sterol in animal tissues. Plant cell membranes contain other kinds of sterols also. Cholesterol (Figure 9) is an alcohol, its -OH group constituting the head group.

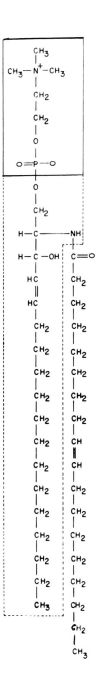

FIGURE 7. Formula of sphingomyelin. See text.

A large variety of lipids are present in different types of cell membranes. Their variety is related not only to the different types of head groups described above but also to the fatty acyl chains. The two acyl chains associated with a head group may be identical or dissimilar. These may be saturated or unsaturated, with up to six double bonds and having an even number of carbon atoms (usually 14 to 24). The occurrence of an even number of carbon atoms in the acyl chains is a natural result of the manner in which they are synthesized. They are built up two carbons at a time from acetate precursors derived from acetyl CoA.

A consequence of the amphipathic nature of the membrane lipids is their ability to form micelles, bilayers, and closed vesicles (Figure 10). The hydrophobic tails of a number of

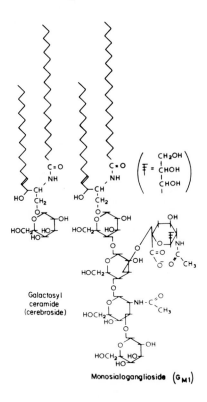

FIGURE 8.   Chemical formulas of a cerebroside and a monoganglioside.

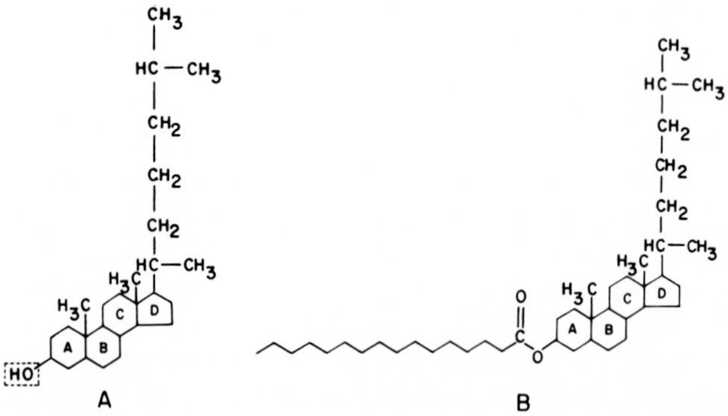

FIGURE 9.   (A) The structure of cholesterol. The boxed hydroxyl is the head group. (B) Cholesterol may be esterified to a fatty acyl chain at the OH.

molecules can move away from the aqueous phase "dissolving" in themselves, thus spontaneously forming micelles. The electrically charged polar head groups of these molecular aggregates face the aqueous medium. A lipid bilayer can form at the junction of two aqueous compartments, the polar groups facing the aqueous phase and the fatty acyl chains forming a continuous hydrocarbon core. Small closed vesicles of such bilayers, designated as liposomes, are readily formed by agitating phospholipids in an aqueous medium. Such liposomes have been studied extensively as membrane models. During their preparation, proteins/ glycoproteins or any other molecules of interest may be included in the aqueous phase

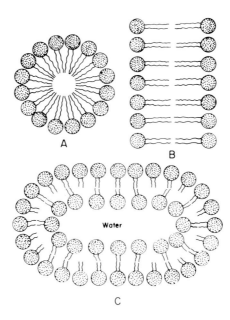

FIGURE 10. Diagram of a micelle (A), a bilayer (B), and a liposome (C).
The polar head groups of lipids are shown by dotted circles.

eventually to study the molecular interactions. The lipid composition also may be varied, permitting studies aimed at elucidation of its consequences to the membrane properties.

Conventionally the lipid molecules are depicted by a circle representing the hydrophilic polar group and two wavy lines representing the fatty acid chains. The wavy depiction suggests double bonds and the kinks resulting therefrom.

b. Proteins

Proteins ranging from molecular weights of 10,000 to several hundred thousand are known to occur in the plasma membrane. In many instances, tissue specific membrane proteins have been demonstrated. Changes in the composition of membrane proteins, the appearance and disappearance of some of them in relation to the functional or differentiated state of cells, are also known in many instances. Many membrane proteins are enzymes and are known to regulate the life processes of the cells. The membrane proteins are undoubtedly responsible for the determination of the "self" and "nonself" features that are recognized by immune mechanisms. It is quite probable that the social behavior of cells is largely determined by the plasma membrane proteins and glycoproteins.

We have seen earlier that the basic structural plan of the plasma membrane is a double lipid layer with proteins associated with it. The location of a protein in the membrane may be peripheral or deep, and this is obviously determined by the nature of the protein. Glycophorin, one of the best known membrane proteins of the human erythrocyte, is a glycoprotein. Functionally it is associated with the determination of the MN system of blood groups. Its primary structure is known.[25] Its NH_2-terminal end projects on the cell surface and the -COOH end is directed towards the cytoplasm. There are 130 amino acid residues in the polypeptide. Of these, a stretch of about 20 amino acids is chiefly hydrophobic. This stretch would approximately correspond to the thickness of a lipid bilayer if the coiling is α-helical. From this, it has been suggested that the hydrophobic region is in the lipid bilayer with the $-NH_2$ terminal end projecting on the cell surface and the -COOH terminal directed towards the cytoplasm. The external hydrophilic domain has 16 oligosaccharide chains, 15 of them attached to serine or threonine through *O*-glycosidic linkage and the 16th to an

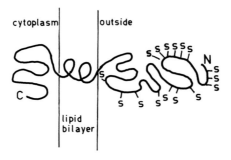

FIGURE 11.   Orientation of glycophorin molecule in the plasma membrane. C and N are, respectively, the C- and N-teminal of the polypeptide. The sixteen amino acid residues on which sugar chains (S) are bound, are located in the part of the polypeptide projecting from the cell surface.

asparagine residue linked through *N*-glycosidic linkage. The glycophorin thus spans the lipid bilayer (Figure 11). Further evidence that glycophorin spans the lipid layer was obtained from studies in which radioiodinated glycophorin was obtained from intact red blood cells and from "leaky" membrane vesicles. It may be expected that in the latter case, lactoperoxidase can reach the tyrosine residues of the proteins projecting on the cytoplasmic face in addition to those on the cell surface. A comparison of the glycophorin labeled in the intact cells and "leaky" membrane vesicles indicated that its C-terminus projects towards the cytoplasm.[26]

Several other proteins of the plasma membrane have been studied. The histocompatibility antigens of mouse and human cells are similar to glycophorin in spanning the lipid layer and having the N- and C-terminal ends projecting on the outer surface and the cytoplasmic side, respectively. However, they differ from glycophorin in having a second lighter polypeptide chain associated with the major polypeptide. The light chain known as β_2-microglobulin, is noncovalently bound to the transmembrane heavier component on the part projecting on the cell surface. Several hydrolases associated with the microvilli of kidney and intestinal epithelial brush borders are integral membrane proteins. They have no carbohydrate components. In case of intestinal sucrase, it is known that the N-terminus is projecting on the cytoplasmic face of the membrane, whereas the C-terminus is on the cell surface. In this respect, sucrase differs from glycophorin and the histocompatibility antigens. It, however, resembles the latter in having a second polypeptide subunit on its exposed part at the surface of the cell. So far we have referred to those integral proteins that have a hydrophobic stretch of amino acids so that they can cross the lipid bilayer once. Some proteins are known to cross the bilayer more than once. An integral glycoprotein of red cell membrane, known as Band 3, is an example of membrane proteins that traverse the lipid layer thrice. Human rhodopsin, the purple visual pigment of the retina, consists of 348 amino acid residues in a single polypeptide and traverses the membrane lipid bilayer seven times[27] (see Figure 12).

c. Carbohydrates

Carbohydrates are important constituents of the plasma membrane, especially from the functional point of view. They occur as branched or unbranched oligosaccharides covalently linked to the lipids or proteins. At the light microscopic level, certain specific staining techniques have demonstrated the occurrence of carbohydrates as components of the plasma membrane. Rambourg,[13] using histochemical staining methods, and Benedetti and Emmelot[28] using colloidal iron, have demonstrated that the sialic acid residues are on the outer surface of the plasma membrane. In high resolution electron micrographs, the electron-dense granules were shown to be restricted to the outer leaflets of the plasma membrane fragments. This is a particularly interesting piece of work since the staining was carried out on membrane

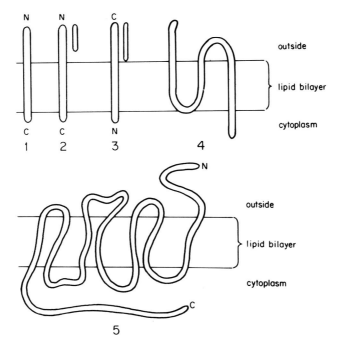

FIGURE 12.   Orientation of the different integral membrane proteins. Proteins traversing the hydrophobic layer once, 1 to 3; thrice, 4; and seven times, 5 are shown. The C- and N-terminals are also indicated. The proteins are exemplified by 1, glycophorin; 2, histocompatibility antigen; 3, sucrase; 4, band-3 protein of erythrocyte membrane, and 5, human rhodopsin.

fragments where the absence of staining of the inner leaflet due to its inaccessibility to the colloidal iron is ruled out. Another interesting observation of Benedetti and Emmelot[28] is that colloidal iron does not stain the membrane at the junctional complexes. Besides colloidal iron, ruthenium red and collodial thorium have also been used for the demonstration of carbohydrates on the plasma membrane surface.[29] The occurrence of specific carbohydrates has been demonstrated by the use of lectins, which are used as membrane probes. We shall describe the properties and uses of lectins later in this chapter.

Nine monosaccharides are found in the glycoproteins and glycolipids of plasma membranes (Figure 13). In glycolipids, the monosaccharide or oligosaccharide chain is linked *O*-glycosidically to the primary hydroxyl group of sphingosine with the fatty acid chains linked to the other hydroxyl groups (Figure 7). Glycoproteins are formed by the covalent binding of the sugar residue on one of the following amino acid residues of a polypeptide chain: asparagine, serine, threonine, hydroxylysine, and hydroxyproline. The sugar bound on these along with the type of linkages are set forth in Table 3. The oligosaccharide chains may be constructed unbranched or branched. There is an important difference between the mechanisms of synthesis of nucleic acids and proteins on the one hand, and of the oligosaccharides on the other. The nucleic acids and polypeptides are synthesized on a template (DNA or RNA), whereas the oligosaccharide chains grow step-wise through the mediation of enzymes known as glycosyl transferases. The monosaccharide is first activated by conversion to the respective nucleotide derivative, which in turn is added to the growing end of the oligosaccharide chain, the transfer being effected by a specific transferase. Each step of addition depends on the completion of the previous step. Thus in the core part of chondroitin sulphate illustrated in Figure 14, the two *Gal* residues are added by the activity of two different transferases. Of the nine monosaccharides illustrated in Figure 13, *N*-acetylneuraminic acid

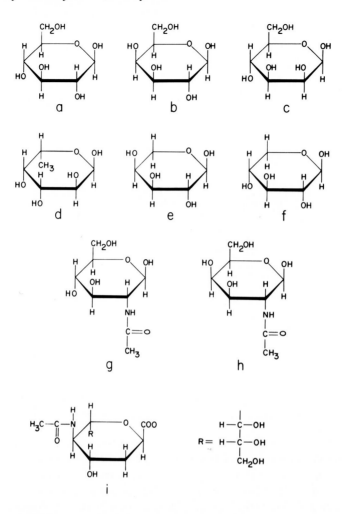

FIGURE 13.   The nine monosaccharides which occur in membrane oligosaccharides. (a) D-glucose; (b) D-galactose; (c) D-mannose; (d) L-fucose; (e) L-arabinose; (f) D-xylose; (g) *N*-acetyl-D-glucosamine; (h) *N*-acetyl-D-galactosamine; (i) *N*-acetylneuraminic acid.

Table 3
OLIGOSACCHARIDE LINKAGES TO PROTEINS

Sugar	Amino acid residue of the polypeptide	Linkage
N-Acetylglucosamine	Asparagine	*N*-Glycosidic
Xylose	Serine	*O*-Glycosidic
N-Acetylgalactosamine	Threonine or serine	*O*-Glycosidic
Galactose	Hydroxylysine	*O*-Glycosidic
Arabinose	Hydroxyproline	*O*-Glycosidic

(NANA) needs a special mention. It is the only one which bears an ionizable -COOH group and is always found as the terminal residue in the oligosaccharide chain. When NANA occurs, the preterminal residue is galactose. An important consequence of the occurrence of NANA residues on the cell surface is the addition of a corresponding number of negative charges, thereby rendering the cell electronegative in physiological pH.

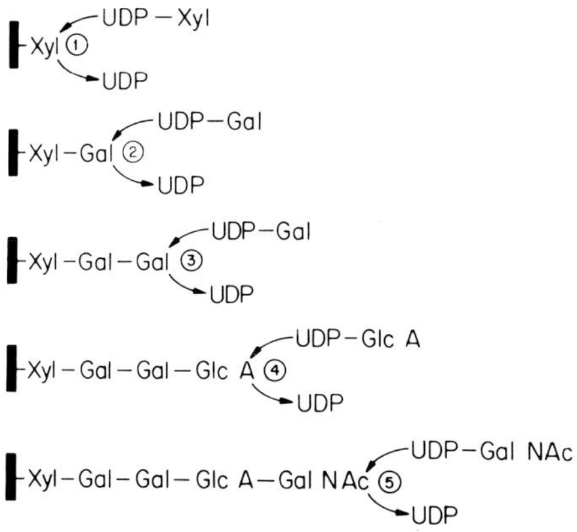

FIGURE 14.   Diagram to illustrate the process of glycosylation by different transferases, 1 to 5. The transferases 2 and 3 are not identical since they transfer galactose on different residues already transferred. The black vertical bar represents a part of the polypeptide on which the sugars are bound. See text.

The occurrence of DNA in the plasma membrane has been reported. Its role in the life of the cell is, however, still in the realm of speculation.[30]

d. *Lectins as Membrane Probes*

Considerable evidence regarding the structure of the plasma membrane has been obtained using lectins as membrane probes. It is therefore appropriate to dwell on this subject even at the risk of appearing to digress.

It was known for a long time that certain plant proteins, especially those obtained from the seeds of various leguminous plants, can bring about agglutination of red blood cells. These proteins were therefore called "hemagglutinins". Recently the term lectin (from Latin *lectus,* the past participle of *legre,* meaning to pick up, choose or select[31]) has come into use. Lectins of animal tissue origin are also known. Some lectins act as mitogens; some are known to be highly toxic. An interesting property of the lectins is that they bind specifically to monosaccharide residues of complex sugars. When whole cells are exposed to lectins, the latter bind to the oligosaccharides of membrane glycoproteins. The binding is noncovalent and can be inhibited competitively by specific monosaccharides. The various lectins have characteristic sugar binding specificities as indicated by competitive inhibition studies. The lectins are at least divalent, i.e., have at least two binding sites, and therefore a single lectin molecule can bind two different cells. Agglutination of cells is brought about by the ramification of cells bound to the lectins. Some lectins commonly used in experimental studies with their sugar specificities are listed in Table 4.

Lectins bind only on the outer surface of the plasma membrane. This is known from experiments in which the lectin, Con A, was conjugated with ferritin, an electron dense material, and allowed to bind on whole erythrocytes or their membrane fragments. Electron micrographs of such cells and membranes show that the lectin binds only on the outer surface. There seems to be nothing repelling the Con A from the protoplasmic surface of the membrane, since ferritin conjugated antibodies against spectrin bind only the inner surface of the membrane and, understandably, not the outer surface. From this it was concluded that the lectin has binding sites only on the cell surface. Another line of evidence leading to the same conclusion is from studies on the binding of fluorescein isothiocyanate-labeled lectins on lymphocytes. The fluorescence-labeled lectin can be visualized on the lymphocyte using a fluorescence microscope. Within 10 min, the fluorescent Con A is found on the cell

Table 4
SUGAR SPECIFICITY OF SOME LECTINS COMMONLY USED IN CELL SURFACE STUDIES

Lectin	Source	Sugar specificity
Concanavalin A (Jackbean lectin; Con A)	*Canavalia einsformis*	α-D-Mannose; α-D-glucose
Wheat germ agglutinin (WGA)	*Triticum valgaris*	N-Acetylglucosamine; sialic acid
Phytohemagglutinin (red kidney bean lectin; PHA)	*Phaseolus vulgaris*	N-Acetyl-D-glucose
Castor bean lectins		
RCA₁	*Ricinus communis*	β-D-Galactose
RCA₂	*Ricinus communis*	D-Galactose; N-acetyl-D-galactose
Abrin (Jequirity bean lectin)	*Abrus precatorius*	D-Galactose
Peanut agglutinin (PNA)	*Arachis hypogea*	D-Galactose; N-acetyl-D-galactose
Horse shoe crab lectin	*Limulus polyphemus*	Sialic acid
Soybean lectin	*Glycine max*	N-Acetylgalactosamine
Lentil lectin	*Lens culinaris*	α-D-Mannose; α-D-glucose
Pea lectin	*Pisum sativum*	Glucose; mannose
Winged pea lectin	*Lotus tetragonolobus*	L-Fucose
Ulex lectin	*Ulex europeus*	L-Fucose

surface. It may also be included in the cell cytoplasm by pinocytosis. However, pinocytosis can be prevented by using inhibitory agents such as antimycin or cyanide. Under these conditions, the fluorescence is found only on the surface of the membrane. No labeling is seen in concurrent control experiments in which the competing sugar, D-mannose, is also added. These experiments suggest that the lectin binding is surface specific. This obviously indicates that the external surface of the cells has carbohydrates.

As mentioned earlier, the binding specificity of lectins can be demonstrated by competitive inhibition of binding by specific sugars. However, the binding specificity in some cases does not seem to be towards simple sugars but to more complex configuration of the oligosaccharides. A lectin extracted from chick embryonic pectoral muscle agglutinates rabbit erythrocytes, and this is inhibited by the disaccharide lactose but not by glucose or galactose.[32,33] It was already known from the work of Allen and Neuburger[34] that more complex sugars are better competitive inhibitors of agglutination. From this it follows that the complex carbohydrates of the cell surface can bind lectins with a very high degree of specificity.

Do the lectins bind only terminal residues or also "core" residues? In case of the *Ricinus* lectins, it has been shown that only terminal residues of the oligosaccharide chains are bound. There are, however, other lectins that can bind to core residues also.[35,36]

The use of lectins as membrane probes was prompted by an accidental discovery of a peculiar difference in the agglutinability of cancer cells and their healthy counterparts. In the early 1960s, cancer researchers had already turned their attention on the plasma membrane, and especially studies on the membrane lipids were being pursued vigorously to examine if tumor cells differed from their normal counterparts by virtue of any differences in the lipids. In this connection, the effects of various lipases on the growth of normal and tumor cells were being investigated. Aub et al.[37] discovered that wheat germ lipase preparations contained an impurity that agglutinated tumor cells preferentially. The lipase activity could be abolished by heating the preparations to 65°C, leaving behind the agglutinin unaffected. It was also shown that the agglutinating principle was nondialyzable and hence a macromolecule, probably a mucopolysaccharide. Subsequently, the wheat germ agglutinin (WGA) was isolated by Burger and Goldberg[38] and shown to be a glycoprotein. The discovery that WGA agglutinates tumor cells specificially gave an impetus to extensive studies on lectins and their action on tumor and normal cells. In general, tumor cells seem to be readily agglutinable in the presence of certain lectins. Using normal and virally transformed mouse

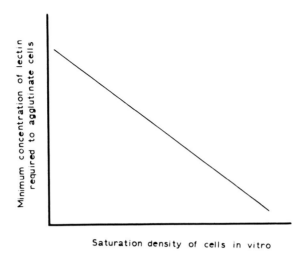

Minimum concentration of lectin required to agglutinate cells

Saturation density of cells in vitro

FIGURE 15.   Diagram showing that the minimum concentration of a lectin such as WGA required to cause agglutination of different types of cells increases with decreasing saturating densities attained by the cell in vitro.

fibroblasts, it has been shown that the greater the malignancy (as measured by limiting densities in culture) the lower is the concentration of WGA required to bring about agglutination (Figure 15). Though this generalization is not without exceptions, it seems to be largely true. In fact, in certain types of cancer (e.g., chronic myeloid leukemia), lectins can be used as diagnostic tools.[39]

The difference between tumor and normal cells reflected in the difference in lectin agglutinability seems to be related to a number of factors including the fluidity of the membrane and the ability of the binding sites to diffuse laterally in the membrane.[35,40] Though lectins bind normal as well as tumor cells, it is only the latter that can agglutinate. At lower temperatures (4 to 15°C), tumor cells bind the lectin, but do not agglutinate, though such cells washed repeatedly to remove unbound lectin agglutinate readily when warmed to 37°C. Normal cells also agglutinate when mildly trypsinized and subsequently exposed to the lectin. The interpretation of these findings is not simple. Recent investigations by Phondke and associates[41-43] have provided evidence to show that Con A binding sites are not uniform. They describe two sets of binding sites on murine lymphocytes, which differ in their mobility characteristics (i.e., capping) at different concentrations of the lectin. They also suggest that the different types of Con A receptors can be used as charcterizing parameters of lymphocyte differentiation.

Another interesting aspect of lectin mediated agglutination is that certain early embryonic cells are readily agglutinated by Con A and other lectins.[44] Agglutinability of embryonic cells seems to be associated with their morphogenetic mobility. Lectin binding of different types of cells will be referred to again in the subsequent chapters of this book.

Studies using lectins have given valuable evidence that the cell surface has carbohydrates that are directly involved in molecular recognition of the cell's environment. Mutual adhesion of the cells and their adhesion to noncellular material is obviously influenced by these carbohydrates. The displacement of lectin "receptors" as a response to lectin binding suggests that similar shifting and rearrangement in surface molecules may be taking place as a result of binding naturally occurring lectin-like molecules in vivo. We shall refer to this subject again in later chapters of this book.

2. Synthesis of Membrane Components

A complex mechanism is involved in the biosynthesis of the membrane components. A

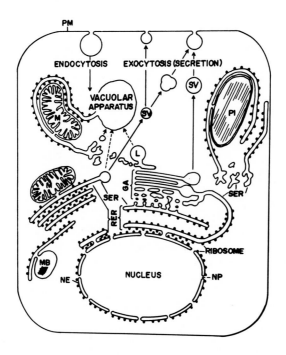

FIGURE 16.   Diagram of the endomembrane system of a cell. GA, Golgi
apparatus; RER, rough endoplasmic reticulum; PM, plasma membrane; SV,
secretory vesicle; L, lysosome; MB, microbody; NE, nuclear envelope; NP,
nuclear pore; M, mitochondrion; Pl, plastid. (From Morré, J. D., *Cell Surface Rev.*, 4, 6, 1977. With permission.)

useful conceptual framework is provided by the notion of the endomembrane system, which
can be described as a functional continuum of membrane types.[45] Structurally the endomembrane system includes the nuclear envelope, rough and smooth endoplasmic reticulum,
Golgi apparatus, and various cytoplasmic vesicles. These structures constitute a complex of
interacting components (Figure 16). Besides the proteins secreted from the cell, the plasma
membrane, vacuole membranes, and lysosomes are considered as the end products of this
system.

Membrane lipids are synthesized in the endoplasmic reticulum. Cytidine activated choline
or ethanolamine combine with diacyl glycerol, forming the corresponding diacyl phosphatidyl
choline or ethanolamine. Elaborate enzyme systems and the biochemical pathways involved
in the synthesis of these and other membrane lipids are described in biochemical texts (see,
e.g., Lehninger).[46] The newly synthesized molecules are, for the most part, inserted into
the lipid bilayer of the endoplasmic reticulum. Golgi body membranes are derived from
the endoplasmic reticulum. Vesicles pinched off from the Golgi get incorporated into the
plasma membrane. Thus lipids are provided for replenishment during the membrane turnover
and for growth. Thus the endomembrane system (Figure 16) participates in the synthesis of
plasma membrane lipids. Membrane lipids may be derived in yet another manner. Plasma
lipoproteins are macromolecular complexes that carry hydrophobic lipids through the aqueous
plasma and tissue fluids to the sites of their deposition or degradation. These complexes are
classified on the basis of the density at which they float during ultracentrifugation. They
are designated as chylomicrons, very low density lipoproteins, low density lipoproteins, and
high density lipoproteins. These protein-lipid complexes have the hydrophilic stretches of
the peptides (apoproteins) exposed to the aqueous medium (Figure 17). Recent work has
shown that the plasma lipoproteins have a definite role to play in maintaining normal cell
membrane lipid composition.[47] Support to such a view has come from pathological conditions

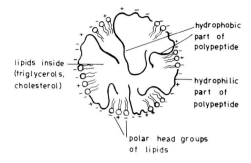

FIGURE 17.   Schematic diagram of a plasma lipoprotein.

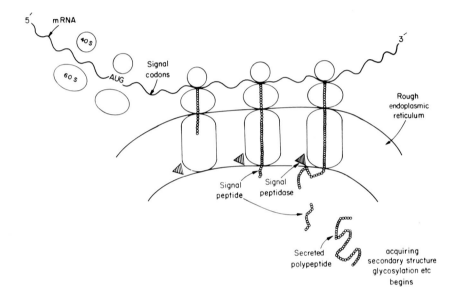

FIGURE 18.   Diagram to illustrate the signal peptide hypothesis.

caused by genetic, hormonal, or dietary disturbance in which alterations in the apoprotein and lipid patterns are observed. Such abnormal plasma lipoproteins are correlated with abnormal membrane lipids.

Much remains to be learned in connection with the synthesis of plasma membrane lipids. For example, little is known about the control of membrane lipid composition and formation of definite lipid domains.

Proteins intended for export from the cell as well as those which become membrane components are synthesized in the cytoplasm and passed through the endomembrane system. It has been proposed that the polypeptides of these proteins are synthesized at the rough endoplasmic reticulum. A short sequence of the messenger RNA for such proteins codes for what is called a *signal peptide* or *leader sequence* that consists of 15 to 30 amino acid residues, most of which are hydrophobic. The N-terminal peptide is able to establish an association of the ribosomes with the membrane and then pass into the lumen of the membrane system. The signal peptide is then removed by a membrane resident peptidase and the remaining polypeptide continues to grow into the lumen of the membrane system (Figure 18).

This mechanism ensures that proteins that are to be exported as secretions or as membrane components are channeled appropriately. Interestingly the peripheral membrane component

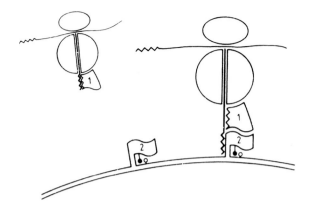

FIGURE 19.   Sequence of events in the translocation of secretory proteins across the endoplasmic reticulum. Translation of the mRNA begins in the cytoplasm on free ribosomes. Translation is blocked by the signal recognition protein (component 1) after the signal sequence has emerged from the ribosomal subunit. Translation is resumed when contact is made with the "docking" protein (component 2). (From Meyer, D. I., Krause, F. and Dobberstein, B. Reprinted by permission from *Nature*, 297, No. 5868, 649. Copyright© 1981, Macmillan Journals Ltd., London.)

proteins such as spectrin are not synthesized in this manner. This seems to make sense as spectrin does not pass *through* the membrane system, but is just associated with the cytoplasmic face of the plasma membrane. Considerable evidence has accumulated to support this signal hypothesis.[48,49] In a cell-free system without the endoplasmic reticulum, it is possible to obtain the synthesis of a secretory protein intact with the signal peptide provided that free ribosomes, the appropriate mRNA, the tRNAs, and the necessary enzymes and precursors are present. If microsomal vesicles are also added to such an in vitro system, the secretory protein without the signal peptide is synthesized. Obviously this is due to the additional mechanisms of processing supplied by the membrane system. For a neat and orderly synthesis of the secretory protein, it is therefore necessary that the growing polypeptide should establish a contact at the proper site (i.e., the endoplasmic reticulum) and then pass through it. Recent work has indicated that a two-step mechanism regulates these events. When rough microsomes are exposed to high salt concentration and elastase, and then added to the cell free system, translocation and the subsequent loss of the signal peptide does not occur. It has been shown[50] that when 60 to 70 amino acid residues of the N-terminus are polymerized on the free ribosome-mRNA complex and the signal peptide has emerged from the large ribosomal subunit, a cytoplasmic protein (250,000 mol wt) selectively stops further translation. This inhibitory protein is known as the *signal recognition protein*. In the microsomal membrane, there is another protein (72,000 mol wt), which has a part of its molecule projecting into the cytoplasm. This is called the *docking protein*. When a ribosomal complex with translation blocked by the signal recognition protein establishes a functional contact with the membrane-located docking protein, the inhibitory block is released. Further translation of the protein and its translocation across the membrane now continues. This two-step regulatory mechanism is illustrated in Figure 19.

An important posttranslational modification of the polypeptides thus synthesized is their glycosylation, resulting in the synthesis of glycoproteins. The Golgi complex (cisternal stack and associated tubules, secretory vesicles, and surrounding zone of exclusion) possesses the terminal enzymes for glycoprotein and glycolipid biosynthesis, consisting of nearly all the glycosylation machinery for adding oligosaccharide sequences to the nascent membrane proteins and lipids. Though glycosylation of polypeptides may start even while they are

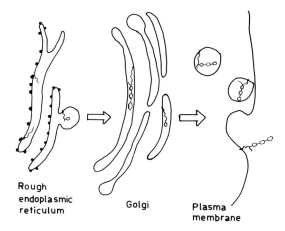

Rough
endoplasmic
reticulum   Golgi   Plasma
membrane

FIGURE 20.   Diagram to show the synthesis of a glycoprotein and its
incorporation into the plasma membrane. The polypeptide (wavy line),
synthesized and translocated across the membrane system of the rough
endoplasmic reticulum, grows and is glycosylated. Addition of sugars
(hexagonal outlines) continues in the Golgi. Vesicles pinched off from the
Golgi migrate to the cell surface and are incorporated into the plasma
membrane.

growing on polysome complexes, and may continue until after their incorporation into the
membrane, the occurrence in Golgi of a large variety of glycosyl transferases suggests that
glycosylation occurs to a great extent in this organelle.

Recent studies have revealed an important difference in the manner in which the carbohydrates are added to the polypeptide chains. The assembly of oligosaccharide chains linked
to asparagine (*N*-glycosidic linkage) follows a pathway in which a series of glycosyl transferases add monosaccharides sequentially on a lipid, pyrophosphoryldolichol. The sugar
donors are nucleotide sugar complexes or other glycosylpyrophosphoryl derivatives. After
the oligosaccharide chain is synthesized, it is transferred *en bloc* from the lipid ''anchor''
to the asparagine residue of a polypeptide which is still growing on the polysomes. In other
words, the process of *N*-glycosylation is essentially a one-step, cotranslational event. On
the other hand, the sugars linked to serine, threonine, hydroxylysine or hydroxyproline
through the *O*-glycosidic linkage are added directly, one by one, to the amino acid residues
without the involvement of a lipid intermediary. This glycosylation occurs chiefly in the
Golgi apparatus. The *N*-glycosidically linked oligosaccharides can ''grow'' further in the
Golgi through the action of respective glycosyl transferases. The recent work on the various
aspects of the synthesis of the glycoproteins is reviewed by Hanover and Lennarz.[51]

It was suggested by Roseman[52] that glycosyl transferases occur at the cell surface and
that glycosylation may occur on the exposed surface of the cell. For a review on the subject
see Pierce et al.[53] On balance, however, it seems that most of the glycosylation occurs in
the Golgi. Regarding the biosynthesis of plasma membrane proteins, a concept of assembly
line has emerged from the evidence. The polypeptides are synthesized in the rough endoplasmic reticulum. After glycosylation as described above, they are released from the Golgi
as a vesicle with the carbohydrates of the glycoproteins directed towards the lumen. The
vesicle, after fusion with the plasma membrane, positions the glycoproteins with the carbohydrates projecting out.[54] This is illustrated in Figure 20.

B. Organization of the Membrane Molecules
1. Membrane Asymmetry and Fluidity
Regional differentiation of the plasma membrane over the cell surface is an observed fact.

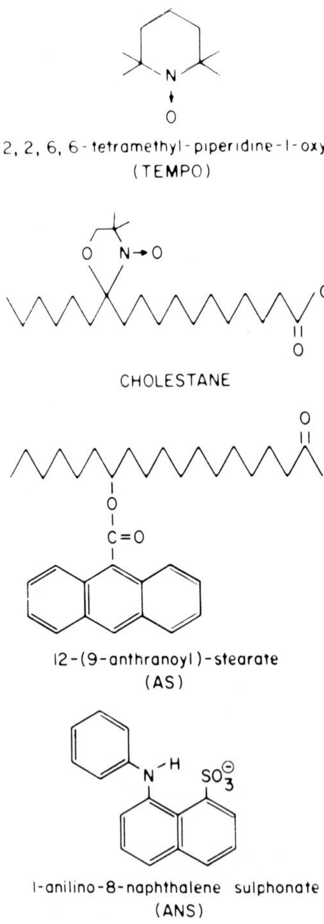

2, 2, 6, 6-tetramethyl-piperidine-1-oxyl
(TEMPO)

CHOLESTANE

12-(9-anthranoyl)-stearate
(AS)

1-anilino-8-naphthalene sulphonate
(ANS)

FIGURE 21.   Fluorescent and spin-label tags for lipids.

The structure of the plasma membrane at the basement membrane and luminal end of epithelial cells is obviously different from each other and from the region where the cell abuts a neighboring epithelial cell. Junctional complexes are specialized structures with intricate molecular architecture. An important functional requirement of the plasma membrane is that it should not act merely as a barrier, but should be selectively permeable. The cell cytoplasm should conserve the pool material and accumulate it against concentration gradients. Besides, the external signals, which are sensed at the cell surface, require that the exposed layer of the membrane be different from the cytoplasmic face. Such asymmetry should be based on the molecular arrangement in the membrane.

The lipids of the outer leaflet are indeed different from those of the inner one. Phospholipase A_2 hydrolyzes two thirds of the membrane phosphatidyl choline from intact erythrocytes, leaving the other glycerophospholipids unaffected. However, if isolated membrane fragments or erythrocyte ghosts are exposed to the enzyme, phosphatidyl choline, phosphatidyl ethanolamine, and phosphatidyl serine are all completely broken down.[55] If the distribution of lipids were symmetrical, i.e., identical in the two layers, one would have expected only 50% breakdown of total phosphatidyl choline from the intact cells. Treatment of intact cells with relatively nonpenetrating, fluorescent, and spin-label reagents for lipids (Figure 21) also indicates that the exposed and cytoplasmic sides of the membrane are not identical.[56] We have already seen that the carbohydrates of the plasma membrane are exclusively on the exposed surface. The observation that phosphatidyl choline can be digested

Table 5
PHASE TRANSITION TEMPERATURES (T$_c$) OF LIPIDS[58]

		Fatty acyl chains[a]		
No.	Phospholipid	1	2	T$_c$, °C
1	Dimyristoyl phosphatidyl choline	14 : 0	14 : 0	23
2	Dipalmitoyl phosphatidyl choline	16 : 0	16 : 0	42
3	Distearoyl phosphatidyl choline	18 : 0	18 : 0	58
4	1-Oleoyl, 2-stearoyl phosphatidyl choline	18 : 1	18 : 0	15
5	1-Stearoyl, 2-oleoyl phosphatidyl choline	18 : 0	18 : 1	3
6	1-Palmitoyl, 2-oleoyl phosphatidyl choline	16 : 0	18 : 1	−5
7	Dioleoyl phosphatidyl choline	18 : 1	18 : 1	−22
8	Dipalmitoyl phosphatidyl ethanolamine	16 : 0	16 : 0	64
9	Dipalmitoyl phosphatidyl serine	16 : 0	16 : 0	53

[a] The fatty acyl chains at the two positions 1 and 2 of the glycerol molecule in the phospholipid (see Figure 6) are indicated separately under 1 and 2. The number of carbon atoms and the number of double bonds are indicated. Thus 18 : 1 is a fatty acyl chain 18 carbon atoms long and with one double bond. It may be noted that the T$_c$ increases with the increasing chain length. The position of the fatty acyl chain also influences the T$_c$ (compare Nos. 4 and 5). Unsaturation lowers the T$_c$ dramatically. The polar group also influences the T$_c$ (compare Nos. 2, 8, and 9).

extensively without affecting the membrane integrity can be interpreted to suggest that the asymmetry of the lipid is stable, with very little exchange occurring between the surface and cytoplasmic layers.

The membrane lipids seem to be in a fluid state in a functionally active cell. The fluidity is determined by several factors. Based on studies using liposomal model membranes, these factors have been identified as (1) the ratio of cholesterol to phospholipids, (2) the degree of unsaturation and length of the fatty acyl chains, (3) the ratio of lecithin to sphingomyelin, and (4) the ratio of lipid to protein. For a detailed discussion, see Shinitzky and Henkart.[57] The transition from a solid to fluid state occurs at a definite temperature (T$_c$, the melting temperature), which is influenced by the degree of saturation of the fatty acid chains and their length.

An accurate method to determine the temperature at which phase transition occurs is differential scanning calorimetry. In this method, the temperature of the lipid sample is raised gradually, and the heat required to raise the temperature is measured. It is observed that at a certain temperature, which is characteristic of the lipid, much more heat has to be applied to raise it further. The temperature at which this latent heat is required is the temperature of phase transition. Saturated fatty acid chains have straight hydrocarbon tails and their abundance results in a higher melting temperature (i.e., greater rigidity). Conversely, unsaturated fatty acids with kinks in the hydrocarbon chains do not permit a crystalline state, and hence the melting temperature is lower (see Table 5). Cholesterol has a dual effect in membrane stability. Plate-like configurations of the cholesterol molecules enter into hydrophobic interactions with neighboring phospholipids, thereby restricting lateral mobility. Hence, the lower the phospholipid to cholesterol ratio, the more rigid is the membrane. Cholesterol may also break up orderly packing of other fatty acids into crystalline structures and therefore has the effect of lowering the melting temperature. Above the melting temperature, however, cholesterol confers rigidity to the membrane structure. In the case of warm-blooded animals, the phase change in the membrane lipids occurs at about 15°C; at temperatures lower than this, the fluidity of the membrane decreases. In view of the fact that the membrane lipids are distributed asymmetrically between the two layers, an additional variable is introduced in the factors determining fluidity.[58] In biological membranes, the

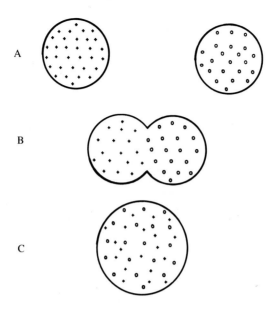

FIGURE 22.   Demonstration of plasma membrane fluidity by the experiment of Frye and Edidin.[60] The different fluorescent labels are depicted
by + and 0. (See text.)

lipid layers are coupled by penetration of a long fatty acid chain of one layer into the other one, and this could also affect fluidity.[59]

The fluid nature of the plasma membrane has been demonstrated by the elegant experiments of Frye and Edidin.[60] When mouse and human cells are fused, using SV virus as an agent promoting cell fusion, the mouse and human antigens are located as separate zones initially and then begin to mingle with each other. In the experiments of Frye and Edidin, the two cell types were labeled with differently colored fluorescent antibodies against their surface antigens. When the cells were just fused, the two colored parts were seen separately, but eventually they intermingled uniformly due to diffusion (Figure 22).

Another line of evidence for membrane fluidity is obtained from what has come to be known as fluorescence recovery after photobleaching. The method is based on photobleaching of a discrete region of the cell surface bearing fluorescently labeled molecules. Photobleaching is achieved by exposing the cell surface to a beam of laser over a micron-sized spot or a lined pattern, thereby destroying the fluorescence. After this, unbleached fluorophores diffuse into the bleached area. For a recent discussion on the subject, the article by Gall and Edelman[61] may be referred to. The speed of recovery is directly related to the diffusion coefficient of the labeled molecules. Using this elegant technique, evidence has been obtained to show that lipids and proteins of the plasma membrane undergo lateral diffusion. The diffusion coefficients for these membrane molecules in artificial lipid bilayers have been found to be in the range of 10^{-8} to 10^{-7} cm^2/sec.

In contrast to the mobility of proteins in the artificial lipid bilayers, diffusion of cell surface glycoproteins has been found to be much slower. This could be due to some restraining forces operating on the membrane glycoproteins *in situ*. For example, diffusion of proteins or glycoproteins anchoring on to a cytoplasmic submembrane structure would restrict its mobility due to diffusion. Indeed, this is the case with some integral proteins of the erythrocyte membrane, whose mobility is restricted by a submembrane matrix of spectrin-actin. In spectrin deficient sphaerocytic mouse erythrocytes, the diffusion rate of these integral proteins is about 50 times higher. Diffusion of integral membrane proteins projecting deep into the cytoplasm can be impeded by submembrane polymer matrix. Diffusion of proteins

MOBILITY RESTRAINT MECHANISMS

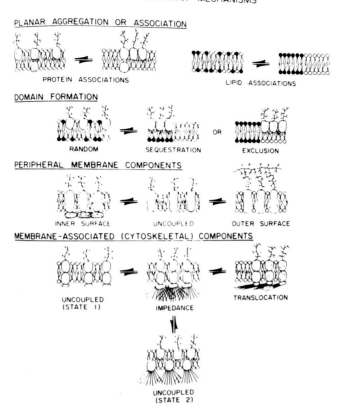

FIGURE 23.   Some probable restraining mechanisms on the lateral mobility of cell surface receptors. (From Nicolson, G. L., *Biochim. Biophys. Acta,* 457, 57, 1976. With permission.)

in cytoplasmic blebs (small globular regions of herniation of the plasma membrane) has been found to be faster than in the unperturbed regions. This accords with the fact that the bleb membrane is devoid of microfilaments unlike the rest of the membrane. The recent studies have been conducted on isolated cells, and it is not yet known how the diffusion of plasma membrane components and their restraint are regulated so as to be relevant to the function of the cells *in situ*. Nicolson[62,63] has depicted the various molecular mechanisms that could control the topographic distribution and associations of cell surface receptors as shown in Figure 23.

The results of fluorescence photobleaching recovery studies have yielded definite information regarding the dynamic state of the molecules constituting the plasma membrane. The lipids also seem to be in a constant thermal motion. Membrane proteins are capable of similar translational movement by diffusion. Indeed, some of the membrane proteins are known to diffuse freely. There are, however, others that are relatively immobile as they seem to be associated with the underlying actin-myosin filaments.[57]

A free diffusion of the lipids and proteins would result in constant changes in the organization of the membrane. Rigidity will, on the other hand, preclude any modulation that may be necessary for the function of the membrane. For functional efficiency, therefore, the two opposite features have to be reconciled to an optimum state. Stabilization of the integral proteins reaching the cytoplasmic boundary is known to be controlled by the cytoskeletal structures. Additional control may be exercised by substances (ligands) on the cell surface. Besides, it is possible that specific associations between proteins and lipids may prevent a truly free diffusion. Are there indeed stoichiometric complexes of proteins and lipids in the plasma membrane? Covalent linking of fatty acids to polypeptides could facilitate stable anchorage of proteins in the lipid bilayer. The occurrence of such acylation has been indicated by Schmidt.[64] If such covalent linking occurs in the membrane, the integral membrane proteins would be surrounded with a definite unchanging set of lipids. However, recent studies using a variety of physical techniques have shown that there is a rapid exchange of free bilayer lipids adjacent to the protein surface.

It is thought by several workers that the lipid exchanges on and off protein surfaces are at the rates of 10^6 to 10^7/sec.[65] In other words, there is no rigid boundary layer of lipids immobilized around the proteins. This evidence, however, does not preclude the existence of lipid domains in the plasma membrane. Regional heterogeneities in the lipids constituting the plasma membrane may exist as indicated by other experimental evidences. Certain fatty acids (oleic, linoleic, and arachidonic) preferably partition into a fluid domain of the lipids, whereas others (elaidic, stearic, and nonadeconic) do so into gel domains. They remain nonesterified in the membrane lipids. At the same time, they alter the phase transition temperature of the lipids with which they are associated. Fatty acids partitioning into the fluid domain decrease the melting temperature of the lipids with which they are associated, whereas those partitioning into the gel phase have the opposite effect. The fatty acids, which decrease the melting temperature, have been shown to decrease the adhesiveness of mouse lymphocytes. They also inhibit the capping of surface immunoglobulins induced by antibodies against them. These alterations in the cells suggest the occurrence of lipid domains in the plasma membrane.[66]

Do specific proteins reside in specific lipid domains? Can the structure and function of the membrane proteins be altered through a controlled alteration in the lipid domain? Can changes in the concentration of membrane proteins alter the bulk fluidity? These questions can be answered if we know more about the plasma membrane architecture. In particular, experimental alteration of the plasma membrane lipids to study its consequences on any recognizable function of the cell would be a fruitful approach. Artificial membrane vesicles are obviously the most attractive models for such studies.

A consequence of the fluid nature of the lipid bilayer is the possibility that the spatial arrangement of the molecules can undergo repeated changes. These changes could constitute the mechanism of modulating the cell surface for functional need. The surface molecules are the obvious receptors for external signals. Any redistribution of the receptors would result in an alteraiton of the cellular response. It is known that the binding of a ligand (such as a lectin) results in redistribution of the receptors. The clusters eventually come together at one pole of the cell (the side close to the Golgi). Other ligands may cause capping at the other pole. There are also ligand-receptor complexes that undergo redistribution on addition of a second ligand (usually an antibody to the first). The mobility of protein/glycoprotein receptors in the fluid layer can thus modify the cell surface in specific ways, depending on the nature of the ligand. Recently it has been shown that the degree of protrusion of the amphipathic proteins into the aqueous phase is determined by the microviscosity of the membrane.[67] This could have important consequences on the function of these proteins, especially if they are the receptors for any signals.

2. Freeze-Fracture Technique and Its Application in the Study of Membrane Architecture

The structure of the plasma membrane revealed by biochemical and other studies has been confirmed and further elaborated by the technique of freeze-fracture. A high resolution replica can be cast on the surface of a hydrated biological material at very low temperatures (below $-100°C$) in a vacuum ($<1.10^{-5}$ mm Hg). The first step in processing the material for the technique consists of freezing it. Obviously, formation of large ice crystals could

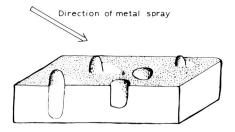

FIGURE 24.   Diagram showing the freeze-fracture surface of the plasma membrane. (See text.)

distort the molecular architecture. Quick freezing minimizes this. This is achieved by using certain fluorocarbons as quenching agents. In addition, free water in the specimen can be reduced by replacing and binding cryoprotectants such as glycerol, dimethyl sulfoxide, ethylene glycol, etc. The use of glutaraldehyde at a low concentration (<1%, 5 to 30 min) as a "fixative" increases membrane permeability to glycerol and stabilizes the membrane architecture. The freezing rate decreases toward the center of the specimen, and therefore, minimizing the size of the material ensures uniform freezing. Holding cells in small droplets and spraying them into liquid propane (at −180°C) has been used as an efficient method. The next step is to fracture the material. Often, the fracture proceeds between the two lipid layers of the plasma membrane. Fracturing is achieved usually with a cold knife (at −196°C). Special techniques have been developed for the fracture of frozen cells in a monolayer. The material can also be frozen between two specimen holders, which when snapped apart, result in its fracture. This provides the two complementary faces of the fracture for study.

Once the specimen is frozen and fractured, it is coated with the vapor of a heavy metal, such as platinum, condensed over its surface. When the beam of platinum vapor falls, forming an acute angle with the fracture surface, it condenses on the elevations and depressions in a characteristic manner. Elevations have a deposit of the metal on the face closer to the source ("windward side"), whereas the other face ("leeward side") remains pale. On the other hand, walls of depressions away from the source get the metal deposit (Figure 24). A faithful replica of the fracture surface in the form of a thin metal film is obtained in this manner. The replica can then be strengthened further by a uniform coat of carbon from a source held at 90° to the surface. Carbon being electron light does not distort the picture of the platinum replica.

The metal coated specimen can be examined by scanning electron microscopy without removing the replica from the fractured surface. It can also be studied using a transmission electron microscope giving much higher resolution. For this purpose, however, the replica has to be thin and be lifted from the surface and placed on a grid. These two are mutually irreconcilable requirements, since the thinnest replica permitting highest resolution is also the flimsiest and most easily damaged while separating from the fractured surface. A suitable thickness for the replica has therefore to be chosen. The specimen below the replica is usually dissolved using a variety of agents such as sulfuric acid, sodium hypochlorite, etc. In transmission electron microscopy, the metal coated areas appear dark, whereas the "shadows" appear light. This point has to be borne in mind while interpreting these "positive images", since the effect is opposite to what one would expect with light: the face of an elevated structure would look bright where light falls and the shadow would be dark. The freeze-fracture pictures published in the literature are positive images.

In a complicated technique such as freeze-fracture, every step can introduce artifacts, and the results have to be evaluated carefully. A discussion on the sources of artifacts and their elimination is, however, beyond the scope of this book. Reference may be made to articles devoted to this.[68,69]

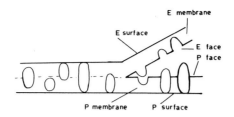

FIGURE 25.   Freeze-fracture terminology. The fracture plane is shown
passing between the two lipid layers of the plasma membrane.

A standard terminology has been suggested for the various surfaces that may be exposed by freeze-fracture.[70] When the fracture passes through the lipid bilayer, it splits the plasma membrane into two half membranes called the P membrane (P for protoplasmic) and E membrane (E for exoplasmic). The exposed surfaces of these are the P face (PF) and E face (EF), respectively (see Figure 25). The free surface of the E membrane (the cell surface *sensu stricto*) is called the E surface and that of the P membrane facing the cytoplasm, the P surface. It may be noted that the A and B faces of the older terminology correspond to the P and E faces, respectively. In addition to the surfaces mentioned above, the fracture may pass through the cytoplasm or the intercellular space.

The freeze-fracture technique has contributed greatly to the understanding of the human erythrocyte plasma membrane structure. The replica of the intercellular space appears smooth, and that of the fracture plane passing through the cytoplasmic mass shows fine granules that have been interpreted to be hemoglobin molecules. The P and E faces of the erythrocyte membrane show a number of particles against a smooth background, those on the PF being more numerous (four to five times) compared with those on the EF. The E surface shows a smooth appearance with numerous globular particles on it. Attempts have been made to identify some of the particles on the freeze-fracture surfaces; glycophorin, which spans the membrane, spectrin and actin, which are on the inner surface (P surface) of the plasma membrane, etc. have been identified in freeze-fracture replicas. It is not intended to give here a detailed account of the freeze-fracture studies. It is to be pointed out, however, that this elegant technique can give visual evidence regarding the molecular architecture of the plasma membrane. Besides erythrocytes, a variety of other cells have been studied.

3. The Singer-Nicolson Model

In the foregoing sections, we have reviewed extant knowledge regarding the composition of the plasma membrane and the dynamic organization of the constituent molecules. An important generalization known as the fluid mosaic model of the plasma membrane proposed by Singer and Nicolson[71] has dominated the field of membrane studies. As mentioned earlier, the experiments of Frye and Edidin[60] demonstrated the fluid nature of the plasma membrane. The fluid mosaic model emphasizes this aspect of the membrane. The model depicts the constituent molecules as shown in Figure 26. A number of observations on living cells vindicate the essential features of the model. Binding of ligands on receptors, their grouping into patches, and eventually capping towards a pole are easily explained by the model. Similarly, the studies using freeze-fracture of cells are in accord with it. An elaborate theoretical discussion on the biophysical and mathematical aspects of membrane dynamics is to be found in a book edited by Perelson et al.[72]

The fluid mosaic model may be considered as an important unifying idea in biology. As with any model, however, minor refinements are necessary to account for new observations made from time to time. In a detailed discussion, Robertson[73] has reviewed the early ideas on the plasma membrane structure and examined the fluid mosaic model in the context of recent knowledge. Depiction of the outer surface of the membrane essentially as a naked

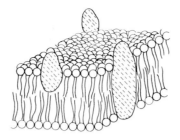

FIGURE 26. Diagram representing the fluid mosiac model as proposed by Singer and Nicolson.[71] The large outlines hatched with broken lines represent the membrane proteins.

surface of lipid head groups is, according to this discussion, an oversimplification. As a matter of fact, most tissue cells are directly in contact with other cells or some cementing substances constituting the extracellular matrix. Thus in vivo, both faces of the bilayer are covered by proteins, glycoproteins, or other soluble substances. Perhaps even the "free" cells such as erythrocytes and the various leucocytes do not exhibit bald stretches of lipid head groups. Support for this idea is provided by the observation that lipase action on whole erythrocytes and ghosts is very different, the intact cell surface being much less susceptible than implied by the membrane model. It is thus likely that the exposed surface of the membrane is largely proteinaceous as depicted in the Danielli-Davson model (Figure 1). A functionally normal cell obviously depends on some external components also. In such a view, these external proteins, aptly called "peripheral proteins", need not be considered strictly as a part of the membrane structure, though of course they are important for the living cell. It is well-known that cells in vitro show very poor adhesion and growth unless the media are supplemented with some proteins usually derived from animal sera. Though the exact role of the serum proteins is not clearly understood, it is likely that they occupy the surface of the cells and make them close to "normal". It is known from a variety of studies that cell surface receptor occupancy is able to modulate internal activities. The serum proteins presumably play a role akin to that performed by the peripheral proteins of cells in vivo. On balance, however, it may be said that the fluid mosaic model has been a very useful generalization.

III. BIOPHYSICAL PROPERTIES OF THE CELL SURFACE

For a developmental biologist, interest in the cell surface is generated by the hope that understanding developmental phenomena could be placed on a more rational foundation after taking into account the molecular interactions involved. During development, cells change their shapes, migrate from one location to another apparently seeking definite target sites, arrange themselves into characteristic patterns, and communicate with each other. The physicochemical properties of the cell surface when fully understood should elucidate these phenomena. The characteristic molecular architecture of the plasma membrane confers on it certain biophysical properties, the most important among them arising from the large number of different ionizable groups (Table 6). The individual ionogenic groups are of course influenced by the surrounding medium, but in addition, are also subject to changes by alterations in the topography of neighboring ionogenic groups of the molecules.[74] Suppression of ionization of certain species can occur as a consequence of interactions with other groups in the vicinity. A complex interaction among the ionizable groups results in a net surface charge on the cell, which in turn could determine its interactions with its cellular and noncellular environment. A very important generalization that has emerged from early

Table 6
pK$_a$ VALUES OF SOME IONOGENIC GROUPS IN PROTEINS

Ionizable group	pK$_a$ (at 25°C)
Phosphate	2.12
Sialic acid carboxyl	2.6
α – COOH	3.0—3.2
β – COOH (of aspartic acid residue)	3.0—4.7
γ-COOH (of glutamic acid residue)	4.5
Imidazolyl (of histidine residue)	5.6—7.0
α-amino	7.6—8.4
ε-amino (of lysine residue)	9.4—10.6
Thiol (of cysteine residue)	9.1—10.8
Phenolic hydroxyl (tyrosine residue)	9.8—10.4
Guanidinyl (arginine residue)	11.6—12.6

From Sherbet, G. V., *The Biophysical Characterization of the Cell Surface*, Academic Press, London, 1978. With permission.

observations on the electrical properties of intact animal cells is that they bear a net negative charge at physiological pH. A quantitative measure of this property can be obtained by electrophoresis, isoelectric focusing, and phase partition behavior of cells.

A. Cell Electrophoresis

When charged particles suspended in a fluid medium are subjected to a DC potential difference, they migrate towards the pole of the opposite sign. Migration of the particle is influenced by several factors, and in order to draw useful conclusions from the electrophoretic mobilities (i.e., migration velocities), some technical aspects of the experiments ought to be understood.

Electrophoretic mobilities determined under standardized conditions of ionic concentration, pH, and viscosity of the bulk medium can be used as characterizing parameters of cells. A variety of instruments have been used for the determination of the electrophoretic mobilities of cells. The most widely used instrument consists of a U-shaped glass tube, the horizontal part of which is a capillary of uniform dimensions, usually 2 to 3 mm in diameter. On the side of this horizontal part is a region of flat surface, which permits viewing the cells by a horizontal microscope. The cell suspension is held in the U-shaped tube and the electrodes are connected to a DC power supply (Figure 27). The U-shaped tube is immersed in a water bath to maintain constant temperature (not shown in the figure). There is also a vertical version of the apparatus.[75]

Electrophoretic migration of the cells is influenced not only by the potential, but also by the electro-osmotic currents caused in the electrolyte medium surrounding the cells. The glass wall itself acts as a charged surface and the fluid flows towards the anode. The flow is fastest close to the wall and decreases towards the center. A compensating flow of the liquid at the center is in the opposite direction (Figure 28). This opposite flow is fastest in the center and gradually decreases towards the periphery. On each radius of the cylinder, there is a point where the liquid is not moving in any direction. This cylindrical laminar layer of the liquid is called the stationary layer or level. It can be shown that the stationary level is at $0.707\,r$ from the center or $0.293\,r$ from the wall. (r is the radius. The volumes of the liquid flowing in the opposite directions are equal. Hence the distance along each radius, which divides the cylindrical space into equal volumes, corresponds to the stationary level. For a detailed discussion see Sherbet).[76] At the stationary layer, the migration of cells is not influenced by the electro-osmotic currents, but only by the electrophoretic pull. The

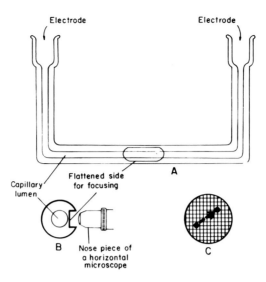

FIGURE 27.   (A) Diagram of a capillary cell electrophoresis tube; (B) the flattened surface shown in section; (C) the apparent path of a cell during electrophoresis.

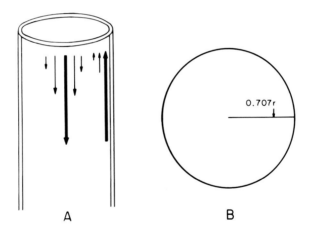

FIGURE 28.   The stationary level in an electrophoresis capillary tube. (A) A tube showing electro-osmotic currents of the saline held in it. The decreasing length and thickness of the arrows depicts diminishing velocities of the flow. (B) A tube in cross section showing the point on a radius where the opposing electro-osmotic currents decrease to zero.

stationary layer of the tube can be determined accurately and the microscope focused to view the cells in this layer during their migration (Figure 27B).

If cells are suspended in saline, they sediment during electrophoresis. The cells thus appear to move in a direction, as shown in Figure 27C. Sedimentation can be prevented by taking an appropriate dense medium, which, however, will correspondingly retard the cell.

In spite of the rather exacting technical precision required as described above, accurate determinations of electrophoretic mobilities of a very large variety of cells have been made yielding valuable information. Cells differ in their electrophoretic mobilities, indicating differences in the net surface charge. Experimental alteration of the cell surface is also expressed as altered electrophoretic mobility (Table 7).[79]

Table 7
ELECTROPHORETIC MOBILITIES OF HUMAN AND MOUSE PERIPHERAL BLOOD CELLS AFTER TREATMENT WITH N-ACETYL NEURAMINIDASE AND RIBONUCLEASE

Corrected to viscosity of water at 25°C

Cell type	Treatment	Electrophoretic mobility (μmsec^{-1}/Vcm^{-1})	% Reduction in mobility, HBSS, v treated	p-value from student's t-test
Human cells				
Lymphocytes	HBSS[a]	0.874 ± 0.015 (180)[b]	—	—
	Neuraminidase	0.505 ± 0.016 (132)	42.4	$p < 0.01$
	Ribonuclease	0.764 ± 0.013 (179)	12.6	$p < 0.01$
Polymorphs	HBSS	0.799 ± 0.20 (130)	—	—
	Neuraminidase	0.451 ± 0.026 (74)	43.6	$p < 0.01$
	Ribonuclease	0.727 ± 0.021 (104)	9.0	$0.02 > p > 0.01$
Monocytes	HBSS	0.572 ± 0.020 (43)	—	—
	Neuraminidase	0.465 ± 0.019 (48)	18.7	$p < 0.01$
	Ribonuclease	0.530 ± 0.022 (51)	7.3	$0.1 > p > 0.05$
Platelets	HBSS	0.958 ± 0.027 (53)	—	—
	Neuraminidase	0.422 ± 0.014 (50)	55.9	$p < 0.01$
	Ribonuclease	0.952 ± 0.030 (51)	0.1	$p > 0.5$
Erythrocytes	HBSS	1.092 ± 0.023 (57)	—	—
	Neuraminidase	0.381 ± 0.018 (53)	65.1	$p < 0.01$
	Ribonuclease	1.074 ± 0.020 (52)	1.7	$p > 0.5$
Mouse Cells				
Peritoneal exudate (Macrophages)	HBSS	0.776 ± 0.025 (77)	—	—
	Neuraminidase	0.629 ± 0.032 (53)	18.9	$p < 0.01$
	Ribonuclease	0.709 ± 0.036 (54)	8.6	$0.17 > p > 0.05$
Thymus cells	HBSS	0.887 ± 0.014 (116)	—	—
	Neuraminidase	0.705 ± 0.015 (113)	20.5	$p < 0.01$
	Ribonuclease	0.828 ± 0.018 (115)	6.7	$p < 0.01$
Liver cells	HBSS	0.836 ± 0.018 (263)	—	—
	Neuraminidase	0.812 ± 0.027 (105)	2.6	$p < 0.05$
	Ribonuclease	0.735 ± 0.019 (100)	11.9	$p < 0.01$

S 37 ascites	HBSS	0.995 ± 0.022 (70)	—	—
	Neuraminidase	0.547 ± 0.045 (31)	45.1	$p < 0.01$
	Ribonuclease	0.818 ± 0.020 (69)	17.8	$p < 0.01$
Erythrocytes	HBSS	1.191 ± 0.031 (50)	—	—
	Neuraminidase	0.932 ± 0.031 (51)	21.7	$p < 0.01$
	Ribonuclease	1.210 ± 0.033 (54)	+1.0	$p > 0.5$

[a] HBSS, Hanks balanced salt solution.
[b] Number of cells observed are given in parentheses.

From Mayhew, F. and Weiss, L., *Exp. Cell Res.*, 50, 441, 1968. With permission.

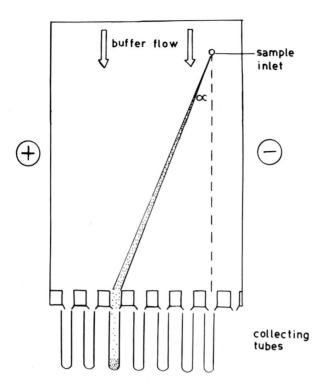

FIGURE 29.   Diagram to illustrate the principle of free-flow electrophoresis. See the text. The discontinuous vertical line represents the path
followed during electrophoresis by particles that bear no net charge. Negatively charged cells are deflected through an angle α.

The electrophoretic method described above is analytical. It is not suitable for separation
of cell populations according to differences in electrophoretic mobilities. Recently a preparative method has been described. The principle of the method is somewhat different. The
apparatus consists of a narrow chamber through which a uniform laminar flow of the
suspending medium is maintained. The thickness of the flowing sheet of liquid is 0.5 to 1.0
mm. Cells are introduced continuously at one point (see Figure 29). The opposite walls of
the chamber act as electrodes. During the flow of the cells along with the suspending medium,
they are also attracted towards the wall due to electrophoresis. The outlets at the bottom of
the apparatus collect the cells. The displacement of the cells (or of a particle with a net
charge) from the path of flow is due to the electrophoretic migration. If the sedimentation
velocity of the cells is negligible in comparison with the flow of the liquid, the electrophoretic
migration is given by the relation

$$\tan \alpha = v\, i / q\, Pw \tag{1}$$

where α is the angle between the vertical course of an uncharged particle and the deflected
diagonal course of the charged particle, v is the electrophoretic mobility of the cell, i is the
current, q is the cross section of the chamber, P is the specific conductivity of the medium,
and w is the velocity of the laminar flow. (The above equation is based on many assumptions
and the paper of Hannig[77] should be referred to for further details.) More recent information
may be obtained from published literature.[78-80] Applications of this technique include separation of macromolecules, whole viable cells, and cell organelles. In particular, separation
of T and B cell enriched populations, separation of renin active kidney cells, and plasma

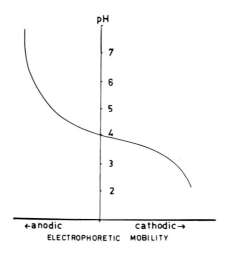

FIGURE 30.   Relationship between electrophoretic mobility of cells and the pH of the bulk medium. The pH scale where the line intersects it corresponds to the isoelectric point. At much higher pH, there is no change in the mobility with corresponding increase of the pH since practically all the anionic groups are in the ionized state. Similarly, below a certain pH there is no increase in mobility with decreasing pH. Here practically all the cationic groups are in the ionized state.

membrane vesicles with the cytoplasmic or outer surface exposed to the medium has been achieved.[81,82]

B. Isoelectric Focusing of Cells

We have already alluded to the fact that the cell surface has a number of ionizable groups (Table 6). The cell as a whole is therefore amphoteric (zwitterionic). The net charge on the cell surface would depend on the pH of the surrounding medium. At physiological pH, the cell has a net negative charge. At very low pH, the net charge is positive. There is a certain pH between these values at which the cell would have no net charge since the positive and negative charges would be equal in number. This pH is called isoelectric pH or isoelectric point (pI). When there is no net charge, the electrophoretic mobility of the cell will be zero. If the electrophoretic mobilities of a cell at different pH values (both above and below the pI) are determined, they can be plotted as shown in Figure 30. Conventionally, mobilities towards the anode are taken as negative. Thus at pH higher than the pI, the mobilities are negative, and at pH lower than the pI, they are positive. From such a plot, the value on the pH scale corresponding to zero mobility is taken as the pI. It must be noted that the pI is not an immutable property, since it is influenced greatly by the ionic concentration and temperature of the environment. However, if accurate values are determined under a standard set of experimental conditions, they can yield useful information regarding the nature of ionogenic groups at the cell surface.

In actual practice, however, obtaining data on electrophoretic mobilities at different pH is both tedious and time consuming. Most of the information on the pI of cells obtained from such studies is restricted to the erythrocytes. From the data given in Table 8, Furchgott and Ponder[83] have calculated the isoelectric point of human erythrocytes to be 1.7. A direct determination of the cell pI is now possible. The method is not simple electrophoresis. If a pH gradient (spanning values both above and below the pI of cells) can be obtained in a vertical column and the cells are dispersed uniformly (as shown in Figure 31) or in any other manner, they bear either a net negative or a positive charge depending on the pH of the surrounding medium. Now if a DC potential is applied, the cells will be repelled from or attracted towards the electrodes, depending on the net charge of the cell. In the course

Table 8
ELECTROPHORETIC MOBILITY OF
HUMAN RED BLOOD CELLS AT VARIOUS
pH LEVELS AND CONSTANT IONIC
STRENGTH OF 0.172 *M* AT 25°C

pH	Buffer system used with NaCl solution to adjust pH	Mobility (μm/sec/V/cm)
10.29	M/10 NaOH—Glycine—NaCl	−1.06
8.35	M/10 NaOH—Glycine—NaCl	−1.06
7.88	M/15 $Na_2 HPO_4$—KH_2PO_4	−1.07
7.32	M/15 Na_2HPO_4—KH_2PO_4	−1.04
5.88	M/15 Na_2HPO_4—KH_2PO_4	−0.98
5.87	M/10 HAc—NaAc	−0.99
4.70	M/10 HAc—NaAc	−0.93
3.86	N/10 HAc—NaAc	−0.83
3.53	M/10 HCl—Glycine—NaCl	−0.80
2.90	M/10 HCl—Glycine—NaCl	−0.62
2.22	M/10 HCl—Glycine—NaCl	−0.34
1.73	0.13 M HCl	0.0

From Furchgott, R. F. and Ponder, E., *J. Gen. Physiol.*, 24, 447, 1941. With permission.

of migration, the cells pass through the medium with gradually varying pH values and would correspondingly change their charge. Eventually when they come to a pH environment equal to the pI, they stop migrating towards the electrodes. Thus cells of the same pI value focus as a band in the column. At this point, the column can be drained into fractions of equal volume. The pH of each fraction and the number of cells are plotted as in Figure 31.

Success of the experiment as described above would depend on how efficiently the pH gradient is generated and maintained. Besides, the application of DC potential introduces two problems: first, it produces bubbles at the electrodes, and second, the current passing through the column produces joule heat that disturbs the pH gradient by causing turbulence currents. The first problem is solved by constructing a suitable apparatus such as the one shown in Figure 32 in which the lower electrode is isolated into a separate chamber. The apparatus shown in Figure 32 also provides an efficient cooling system when a suitable coolant is circulated through it. In developing a method of isoelectric focusing (IEF), the generation of a pH gradient was the most important problem, and it was successfully solved by the synthesis of polyamino-polycarboxylic acids[83,84] (see Figure 33) differing in their pK values. When subjected to a potential difference, they arrange themselves into a "natural" pH gradient. Mixtures of such synthetic molecules variously known as Ampholines® (LKB, Sweden) Pharmalytes® (Pharmacia, Sweden), etc. are now commercially available. Subsequently, the use of certain zwitterionic buffers was introduced, dispensing with the need for the more complex mixtures such as Ampholines.®[85] The buffers can be mixed in such a manner as to give an approximately linear pH gradient, thereby reducing the time taken for the experiments.[86] The pH gradient generated by any of the methods is generally superimposed by a suitable density gradient (ficoll, sucrose, glycerol, polyethylene glycol, and other substances in decreasing order of preference) in order to prevent mixing by turbulence. Heavy water can also be used. A mixture of ficoll and sucrose has been recommended by Boltz et al.[87] This has the advantage of maintaining the osmotic conditions uniformly throughout the column, thereby providing a milieu least damaging to the cells. A detailed discussion of the instrumentation and other aspects of the technique is beyond the scope of this text. Reference may be made to the following books: Sherbet[76] and Arbuthnott and Beeley.[88]

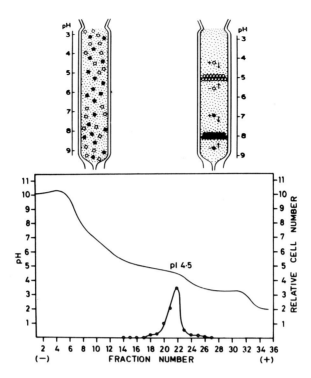

FIGURE 31.   Isoelectric focusing. Particles such as proteins, cell organelles, or whole cells bearing positively and negatively ionizable groups exposed to the medium can be separated by isoelectric focusing. The column at the upper left shows the initial random distribution of hypothetical particles introduced into a pH gradient. When a DC potential is applied, the particles are repelled or attracted towards the electrodes according to the net charge on them. As they pass through the medium, the net charge varies according to the ambient pH. Eventually the particles are focused as separate bands at a pH of the medium where they bear no net charge (upper right). The lower half of the figure shows the profile obtained in an actual experiment. The gradually falling line represents pH. Chick embryonic liver cells are eluted as a more or less symmetrical peak around pH 4.5. (From Rao, K. V., Grover, A., and Beohar, P. C., *Prog. Clin. Biol. Res.,* 151, 345, 1984. With permission.)

One of the earliest attempts at isoelectric focusing of cells was that of Ave et al.[89] The pH gradients generated in their experiments were not satisfactory, presumably due to the high conductivity of the column. Considerable improvements have been made in the technique[90] using Ampholines® and zwitterionic buffers.[91,92] Isoelectric focusing of cells can be used as a method of characterization of cells and also as a separation technique.[76,93] A brief list of cells that have been subjected to IEF along with the experimental conditions has been given in Table 9. From this it will be clear that the method can be applied to a wide variety of cells in studying the plasma membrane and its molecular architecture.

C. Electrokinetic Behavior of Cells: Some Theoretical Considerations

In order to gain an insight into the mechanisms of the electrokinetic behavior of cells, a few physicochemical principles ought to be understood. The brief account of these given here may be supplemented by reference to other literature on the subject.[1,76] In general, the electrophoretic mobility of a particle, such as a cell, is due the ionogenic groups at the surface. All animal cells studied so far have been found to possess a net negative charge on the surface when they are suspended in media at physiological pH. The rate of migration (velocity or electrophoretic mobility) would be directly proportional to the number of net charges on the surface. However, this relation will hold good only if there are no ions in the fluid medium. In practice, however, the cells are suspended in ionic media. After the cells are introduced into an ionic medium, a number of positively charged ions would

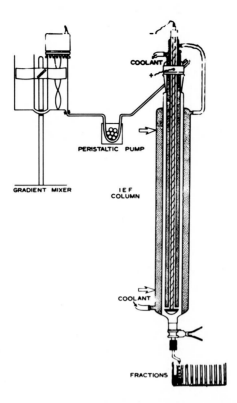

FIGURE 32.. Design of the isoelectric focusing column supplied by LKB-Produkter AB, Bromma, Sweden. The density gradient is generated using the gradient mixer and transferred into the column using a peristaltic pump. The part of the column shown dotted, contains a circulating coolant. The density-pH gradient is held in the annular part, which is not shaded.

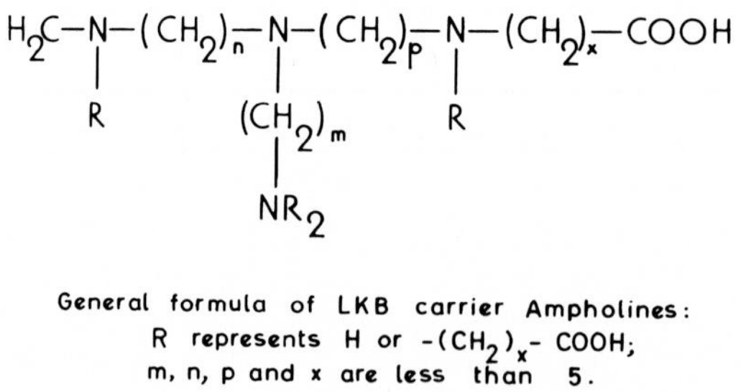

$$H_2C-N-(CH_2)_n-N-(CH_2)_p-N-(CH_2)_x-COOH$$

with R, $(CH_2)_m$, R substituents, and NR_2

General formula of LKB carrier Ampholines:
R represents H or $-(CH_2)_x- COOH$;
m, n, p and x are less than 5.

FIGURE 33. The general formula of carrier ampholines supplied by LKB Produkter AB, Sweden.

accumulate around the surface negative charges. This has the effect of decreasing the surface potential of a cell.

Let us consider a cell (assumed, albeit artificially, to be spherical and having radius r) in an ionic medium such as isotonic saline at physiological pH. The cell bears Q negative charges. Cations in the medium tend to surround the cell due to attractive forces. The counterions nearest the surface are sufficiently large, and their nuclei are therefore unable to reach the cell surface. A monolayer of positively charged ions may therefore be insufficient to neutralize the surface, and so additional layers build up. These ions will also attract free anions in the medium. A cloud of ions is thus built around the cell. Between the ionic cloud

Table 9
ISOELECTRIC POINTS OF SOME CELLS

Cell type	pI[a]	pH Gradient generated by	Ref.
Embryonic Cells			
Chick neural retina (stage 33)[b]	4.64 ± 0.10	Zwitterionic buffers	92
Chick myoblasts from leg (stage 34)	4.23 ± 0.10	Zwitterionic buffers	92
Chick brain (stage 26)	4.36 ± 0.05	Zwitterionic buffers	92
Chick kidney (stage 33)	4.35 ± 0.08	Zwitterionic buffers	92
Chick liver (stage 34)	4.60 ± 0.12	Zwitterionic buffers	92
Rat brain (13 days gestation)	4.96 ± 0.04	Zwitterionic buffers	92
Rat kidney (17 days gestation)	4.75 ± 0.13	Zwitterionic buffers	92
Established Cell Lines			
HFS (from a human fibrosarcoma)	4.6[c]	Zwitterionic buffers	92
HeLa	6.85 ± 0.11	Ampholines	90
Polyoma transformed BHK-21 hamster kidney fibroblasts	6.4 ± 0.01	Ampholines	90
Yoshida ascites sarcoma	6.35 ± 0.19	Ampholines	90
Ehrlich ascites	5.6[c]	Ampholines	90
Miscellaneous			
Chick bursa lymphocytes (1 day after hatching)	3.18 ± 0.18	Zwitterionic buffers	93
Chick thymus lymphocytes (1 day after hatching)	4.52 ± 0.03	Zwitterionic buffers	93
Cynechococcus (a unicellular blue green alga)	4.25[c]	Zwitterionic buffers	94
Chlamydomonas reinharditii (a cell wall-less mutant)	3.00[c]	Zwitterionic buffers	94
Rat peritoneal leucocytes	4.20[c]	Ampholines	87
Rabbit spermatozoa, ejaculated	6.80	Ampholines	87
E. Coli	6.10	Ampholines	76

[a] Data given are mean ± SE.
[b] Chick embryo stages according to Hamburger and Hamilton.[102]
[c] Where data are insufficient, mean of two or three determinations is given.

and the cell surface there would be a space of practically no ions (Figure 34). Though this space is in fact very small, note that it has been exaggerated in the figure. A few positive ions possibly do penetrate this layer as depicted in the figure. When the cell is subjected to a potential difference, it migrates along with this cloud, and the potential at the surface of shear (the slipping plane) is known as ζ potential. It is the ζ potential that determines the electrophoretic mobility of the cell. Away from the surface of shear, the potential falls rapidly. We can thus visualize three layers of potential around the surface: the potential at the surface (ψ, the Nernst potential), the potential in the immediately surrounding zone (the Gouy-Chapman potential), and the potential at the slipping plane (ζ potential). These three levels are shown in Figure 34. Under the conditions of electrophoresis, the net potential at the surface of the cell moving against the bulk medium (i.e., the cell and the ions moving with it) is the ζ potential. This is related to the electrophoretic mobility, *v*, as follows:

$$v = \frac{\zeta D}{4 \pi \eta} \qquad (2)$$

where η is the viscosity of the medium and *D* is the dielectric constant of water. Considering the cell of radius *r*, and a net negative charge potential, $-Q$, at the surface and the layer of counterions of thickness *d* and of potential $+Q$, the ζ potential can be shown to be

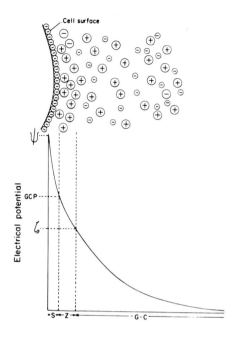

FIGURE 34.   Graphic illustration of the cell surface charge distribution and the potential arising from it. The upper part of the figure depicts a portion of the cell surface bearing a large number of negatively charged groups on it. The surrounding medium contains both positively and negatively charged ions, the former in greater density nearer the cell surface. Away from the cell surface the two types of ions tend to be equal in distribution. The lower part of the figure depicts the different electrical layers and potentials: G-C, Gouy-Chapman diffuse layer; GCP, Gouy-Chapman potential; S, Stern layer; Z, zone of shear; ψ, Nernst potential at the cell surface; ζ, zeta potential. (Adapted from Sherbet, G. V., *The Biophysical Characterization of the Cell Surface*, Academic Press, London, 1978.)

$$\zeta = \frac{Q}{D(r + d)} - \frac{Q}{D\,r}$$

$$\therefore \zeta = \frac{Q}{D\,r} \times \frac{d}{r + d} \tag{3}$$

In case of animal cells, r is much greater than d, and we can put $r = r + d$. Equation 3 can therefore be written as

$$\zeta = \frac{-Q\,d}{D\,r^2} \tag{4}$$

Now substituting this value of ζ, Equation 2 becomes

$$v = \frac{-Q\,d}{4\,\pi\,r^2\,\eta} \tag{5}$$

In the above relation, it can be seen that the electrophoretic mobility is directly proportional to the charge density (number of negative charges per unit area of the spherical surface which is expressed by $-Q/4\,\pi\,r^2$) and inversely proportional to η, the viscosity of the medium. Equation 5 can be written in another form

$$-v = \frac{\sigma}{K\,\eta} \tag{6}$$

where σ is the net negative charge density and K is the inverse of d, which is called the Debye-Hückel function. It may also be noted that the expression v, i.e., the electrophoretic mobility, has a negative sign indicating that the mobility is towards the anode. In the experiments using the horizontal or vertical capillary apparatus, the velocity is expressed as $\mu m\ sec^{-1}/Vcm^{-1}$. The expression Vcm^{-1} represents the DC potential applied. The influence of viscosity may be maintained constant by using media of the same composition in all experiments. Another way of expressing the mobility is using the Tiselius unit, which is equal to $10^{-5}\ \mu m\ sec^{-1}/Vcm^{-1}$ with the minus sign indicating anodic mobility. Electrophoretic mobility can also be expressed to indicate the migrating velocity of a cell in terms of the migrating velocity of an arbitrarily chosen cell type such as glutaraldehyde fixed human erythrocytes.

D. Phase Partition of Cells in Aqueous Two-Phase System

A complex surface property is exhibited in different affinities of cells to media of different biophysical nature. This is an aspect of fundamental importance and will eventually elucidate cell behavior in development. Partition of cells between immiscible liquids is an expression of such surface properties. When aqueous solutions of two different polymers (e.g., dextran and polyethylene glycol) above certain concentrations are mixed together, they form immiscible two-phase systems, even though water is a common constituent of both phases. If shaken and allowed to settle in a tube, they form two layers, one above the other, the upper one being polyethylene glycol-rich and the lower, dextran-rich. In such a system, if some soluble substances are dispersed, they eventually partition in such a manner that their concentration in the two phases is unequal. Larger "particles" such as cells and membranes also show a similar behavior. The cell property that determines the differential affinity for the two phases is chiefly surface charge, though other molecular interactions are also involved. Phase partition of cells has been developed into an extremely sensitive method for separation and subfractionation of cell populations.

The partition behavior of a substance can be described by the partition coefficient K, which is the ratio between the amounts of the substance in the upper and lower phases.

$$K = \frac{[C_{top}]}{[C_{bottom}]} \tag{7}$$

where the brackets represent molar concentrations. It has been deduced that K may be represented by

$$K = \exp\left[\frac{A\,\lambda}{R\,T}\right] \tag{8}$$

where A is the surface area of the particle, R is the gas constant, T the absolute temperature, and λ is a constant which represents a complex characteristic (chiefly surface charge) of the particle-aqueous phase interactions. It may be noted that there is an exponential relationship between K and the particle attributes A and λ. From this it follows that even small changes in A and λ cause very significant changes in the partition coefficient. Hence phase partition is a sensitive method to characterize the surface properties of particles.

Immiscible aqueous solutions of polymers are obtained only above certain "critical" polymer concentrations. Below the critical concentration, the two phases are miscible. At concentrations higher than the critical, partition of cells decreases. In practice one has to determine the relative concentrations of the polymers in the system so as to obtain optimum partition characteristics.[95] Salts and buffers are included in the polymer solutions so as to protect the cells osmotically. The ions influence the partition characteristics, and hence the ionic composition may be manipulated so as to increase partition. An additional effect of the ions in the system is that they are also distributed unequally between the two phases generating an electrostatic potential difference. This factor also influences separation. A system containing 5% (w/w) Dextran® T 500 (molecular weight ≈500,000) and 4% (w/w) polyethylene glycol (molecular weight ≈4000 or ≈6000) with added buffer salts has been used extensively in cell separation work.

More often than not, the discriminatory ability of the two-phase system is not enough to bring about a clear separation of two components in a mixture. Countercurrent distribution, a method of multiple stage partition, can give improved separations. The following description, which illustrates the principle of countercurrent distribution, is adapted from Sherbet.[76] Let us consider a mixture of two components A and B whose partition coefficients are assumed to be 2.3 and 4, respectively. Let us further assume that initially the two components are in equal amounts, viz., 100 arbitrary units of concentration in the mixture.

When the mixture is dispersed in the two-phase system, agitated and allowed to settle, the concentration of the component A in the upper and lower phases will be 70 and 30 units, respectively. Similarly, substance B is partitioned, resulting in 80 units in the upper phase and 20 in the lower. This is the initial stage of separation (Stage A; see Figure 35). After the two phases are separated and equilibrium established, the upper phase is mixed with fresh lower phase polymer, agitated, and allowed to settle. During this Stage B separation, the two substances again redistribute according to their partition coefficients. Concurrently the lower phase of the Stage A is mixed with fresh upper phase polymer and allowed to equilibrate. During the successive stages of separation, the phases are mixed as shown in Figure 35. At stage H when we have eight tubes, the substances already show a bimodal distribution (Figure 36). If this process is continued further, a complete separation of the two substances can be achieved.

When large particles such as cells are partitioned in the two-phase system, the distribution is between one bulk phase and the interface (Figure 37). The partition coefficient is independent of the total concentration of cells, and the ratio of the volume of upper phase to lower. Partition in such cases is described by the following formula:

$$K = \frac{n_1}{n_i} = \exp\left[\frac{\Delta E}{K T}\right] \tag{9}$$

where K is the partition coefficient, n_1 and n_i are the number of cells in the bulk phase and interface, respectively; E is the work required to transfer one cell from the bulk phase to the interface, K is Boltzmann's constant, and T the absolute temperature. When the cell surface charge is considered, the equation may be written as

$$K \frac{n_1}{n_i} = \exp\left[\frac{A}{K T} (\lambda + \sigma V)\right] \tag{10}$$

in which the various symbols are similar to those in Equation 9, and in addition, A and σ are the surface area and charge density, respectively; V and λ are system parameters, the transfacial electrostatic potential difference, and a convolution of the three relevant interfacial tensions. (σ is anionic and V is of conventional polarity.) The partition ratio, K', which is the fraction of the total cells remaining in the bulk phase after equilibration, is related to K as

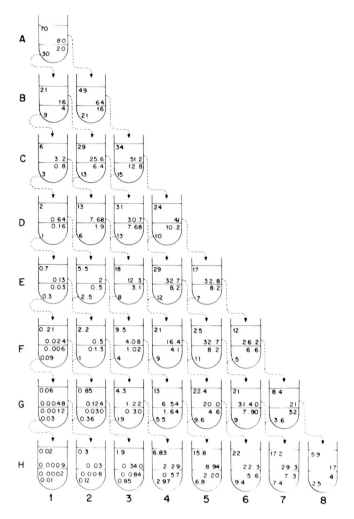

FIGURE 35..   Diagram to explain the method followed for separating substances by the method of countercurrent distribution. Two hypothetical substances of partition coefficients (*k*) 2.3 and 4 are considered. The initial amount of each of the substances is 100 units of concentration. In each tube the numbers on the left represent the amount of the substance with *k* = 2.3; on the right are the corresponding numbers for the other substance with *k* = 4. A-H are successive steps of the experiment. After equilibration, the tubes B1, C1, D1, E1, F1, G1, and H1 receive fresh upper phase polymer to be mixed with the lower phase polymer of the previous step. Similarly the tubes B2, C3, D4, E5, F6, G7, and H8 receive the lower phase polymer. In the other tubes, the lower and upper phases are mixed as indicated by the arrows. When the experiment passes through eight steps there are eight tubes, 1 to 8. (Adapted from Sherbet, G. V., *The Biophysical Characterization of the Cell Surface,* Academic Press, London, 1978.)

$$K' = \frac{K}{1 + K} \quad \text{or} \quad K = \frac{K'}{1 - K'} \tag{11}$$

From the exponential relation beteen K and σ, it follows that

$$\ln\left(\frac{K'}{1 - K'}\right) \alpha \ \sigma \tag{12}$$

We have seen earlier (Equation 5) that the electrophoretic mobility is proportional to the surface charge density. Hence there should be a correlation between ln $(K'/1 - K')$ and electrophoretic mobility. Ballard et al.[96,97] have shown that such a correlation does exist.

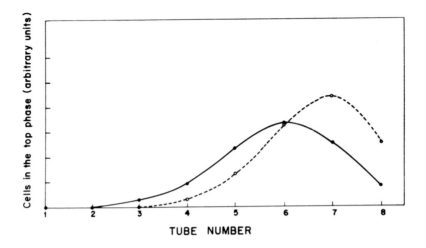

FIGURE 36.   Distribution of two substances in the countercurrent distribution experiment illustrated in Figure 35. The continuous line represents the substance with $k = 2.3$ and the broken line represents the substance with $k = 4$. (Reproduced from Sherbet, G.V., *The Biophysical Characterization of the Cell Surface*, Academic Press, Orlando, 1978. With permission.)

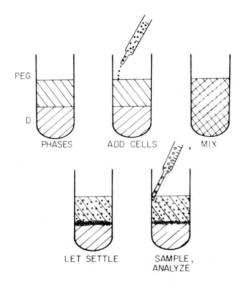

FIGURE 37.   Partition of cells in two-polymer aqueous phases. (From Walter, H., Krob, E. J., and Ascher, G. S., *Exp. Cell Res.*, 55, 279, 1969. With permission.)

Partition in polymer-electrolyte phase systems as described above offers a very powerful technique to characterize the surface properties of whole cells. Though the method has not been used as extensively as cell electrophoresis, it has enough versatility for application in a wide area of cell research. Obviously no particular two-phase system will be suitable for all kinds of cells. Several variables in the experimental procedure, which can influence partition behavior, have been investigated, and the method of arriving at optimal systems for particular needs has been suggested.[96,97] As an example of the application of this technique, the study by Bandyopadhyay[101] may be referred to.

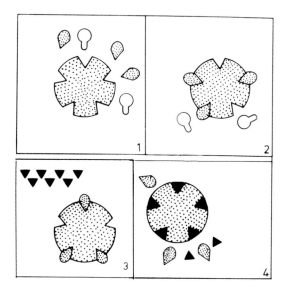

FIGURE 38. The principle of affinity chromatography of cells. 1, A bead of the affinity matrix (large dotted outline) is shown with two types of cells represented by different shapes. 2, The cells that do not bind specifically are eluted with the column buffer. When the specific eluent (dark triangles) is added — 3, the bound cells are dislodged and eluted — 4.

E. Affinity of Cells for Solid Objects

Many of the functions of cells in vivo are thought to be related to their surface molecules. In the animal tissues, cells seem to be specifically attached to other neighboring cells or substances of the intercellular matrix. Under certain circumstances, cells can get dislodged from their place of adhesion and move on to other sites. Conversely, cells wandering randomly may be sequestered by definite anatomical parts. Affinities of cells can be studied experimentally, and the molecular mechanisms underlying the adhesive specificities can be revealed if we can identify the molecules participating in the process.

Affinity chromatography, a technique routinely used in biochemical separations, has been adapted for characterization and separation of cells with respect to their affinity for solid objects. This has been made possible by the development of large sepharose beads (Sepharose® 6 MB, a trade mark product of Pharmacia Fine Chemicals, Sweden) on which a variety of substances (ligands) such as lectins, specific antibodies, and hormones can be immobilized. This is achieved by covalent binding of the primary amino groups of the ligand molecules on CNBr-activated sepharose beads.[98] The beads are much larger (200 to 300 μm diameter) compared to most animal cells. When packed in vertical columns, they leave large interstices through which unbound cells can pass down along with the elution buffer. When cells come in contact with such beads, they bind noncovalently on the specific ligands. Cells having no affinity for the ligands do not bind and are therefore not retained on the beads. Now if a substance which can compete for the binding sites on the ligand molecules is added, the cells are dislodged. In this manner, the method can be used for characterization of cells as well as their separation from a complex population. Application of the competitive eluent in increasing concentrations can fractionate ligand-bound cells into classes of increasing specificity of binding. The principle of affinity chromatography of cells is illustrated in Figure 38.

Some of the early applications of this method may be mentioned briefly. Using α-bungarotoxin immobilized on Sepharoses® 6 MB, a ≈95% enriched fraction of viable and electrically active acetyl choline receptor-bearing neuronal cells could be obtained from sympathetic ganglia.[99] Human peripheral T lymphocytes could be fractionated into subgroups eluted by different concentrations of the competing sugar, *N*-acetylglucosamine.[100] It was shown that the cells separated in this manner differ in their biological properties. In the recent past, this method is finding application in a wider area of biological research.

REFERENCES

1. **Curtis, A. S. G.,** *The Cell Surface: Its Molecular Role in Morphogenesis,* Logos Press, Academic Press, London, 1967.
2. **Overton, E.,** Über die allgemeinen osmotischen Eigenschaften der Zelle, ihre vermutlichen Ursachen und ihre Bedeutung für die Physiologie, *Vjschr. naturf. Ges. Zürich,* 44, 88, 1899.
3. **Görter, E. and Grendel, R.,** On bimolecular layers of lipoids on the chromocytes of blood, *J. Exp. Med.,* 41, 439, 1925.
4. **Cook, G. M. W. and Stoddart, R. W,** *Surface Carbohydrates of the Eukaryotic Cells,* Academic Press, London, 1973, 3.
5. **Danielli, J. F. and Davson, H.,** A contribution to the theory of permeability of thin films, *J. Cell. Comp. Physiol.,* 5, 495, 1935.
6. **Touster, O., Aronson, N. N., Jr., Dulaney, J. T., and Hendrickson, H.,** Isolation of rat liver plasma membranes. Use of nucleotide pyrophosphatase and phosphodiesterase I as marker enzymes, *J. Cell Biol.,* 47, 604, 1970.
7. **Fleischer, S. and Packer, L.,** *Biomembranes, Part A, Methods in Enzymology,* Vol. 31, Academic Press, New York, 1974.
8. **Glick, M. C.,** Isolation of surface membranes from mammalian cells, in *Mammalian Cell Membranes,* Vol. 1, Jamieson, G. A. and Robinson, D. M., Eds., Butterworths, London, 1976, 45.
9. **Beaufy, H. and Amar-Costesec, A.,** Cell fractionation techniques, in *Methods in Membrane Biology,* Vol. 6, Korn, E. D., Ed., Plenum Press, New York, 1976, 1.
10. **Wallach, D. F. H. and Winzler, R. J.,** *Evolving Strategies and Tactics in Membrane Research,* Springer-Verlag, New York, 1974.
11. **Warren, L., Glick, M. C., and Nass, M. K.,** Membranes of animal cells. I. Methods of isolation of the surface membrane, *J. Cell. Physiol.,* 68, 267, 1967.
12. **Wallach, D. F. H. and Schmidt-Ullrich, R.,** Isolation of plasma membrane vesicles from animal cells, in *Methods in Cell Biology,* Vol. 15, Prescott, D. M., Ed., Academic Press, New York, 1977, 235.
13. **Rambourg, A.,** Morphological and histochemical aspects of glycoproteins at the surface of animal cells, *Int. Rev. Cytol.,* 31, 57, 1971.
14. **Maddy, A. H. and Dunn, M. J.,** The solubilization of membranes, in *Biochemical Analysis of Membranes,* Maddy, A. H., Ed., Chapman & Hall, London, 1976, chap. 6.
15. **Dahl, J. L. and Hokin, L. E.,** The sodium-potassium adenosine triphosphate, *Annu. Rev. Biochem.,* 43, 327, 1974.
16. **Hubbard, A. L. and Cohn, Z. A.,** Specific labels for cell surfaces, in *Biochemical Analysis of Membranes,* Maddy, A. H., Ed., Chapman & Hall, London, 1976, chap. 12.
17. **Gahmberg, C. G. and Hakomori, S.,** External labeling of cell surface galactose and galactosamine in glycolipid and glycoprotein of human erythrocytes, *J. Biol. Chem.,* 248, 4311, 1973.
18. **Steck, T. L. and Dawson, G.,** Topographical distribution of complex carbohydrates in the erythrocyte membrane, *J. Biol. Chem.,* 249, 2135, 1974.
19. **Datta, P.,** Labeling of the external surface of hamster and mouse fibroblasts with [^{14}C] sialic acid, *Biochemistry,* 13, 3987, 1974.
20. **Veerkamp, J. H. and Broekhuyse, R. M.,** Techniques for the analysis of membrane lipids, in *Biochemical Analysis of Membranes,* Maddy, A. H., Ed., Chapman & Hall, London, 1976, chap. 8.
21. **Cook, G. M. W.,** Techniques for the analysis of membrane carbohydrates, in *Biochemical Analysis of Membranes,* Maddy, A. H., Ed., Chapman & Hall, London, 1976, chap. 9.
22. **Levine, Y. K.,** Physical studies of membranes, in *Mammalian Cell Membranes,* Vol. 1, Jamieson, G. A. and Robinson, D. M., Eds., Butterworths, London, 1976, chap. 4.
23. **Chapman, D.,** Physicochemical studies of cellular membranes, in *Mammalian Cell Membranes,* Vol. 1, Jamieson, G. A. and Robinson, D. M., Eds., Butterworths, London, 1976, chap. 5.
24. **Smith, R. L. and Oldfield, E.,** Dynamic structure of membranes by deuterium NMR, *Science,* 225, 280, 1984.

25. **Marchesi, V. T., Tillack, T. W., Jackson, R. L., Segrest, J. P., and Scott, R. E.,** Chemical characterization and surface orientation of the major glycoprotein of the human erythrocyte membrane, *Proc. Natl. Acad. Sci. U.S.A.,* 69, 1445, 1972.

26. **Segrest, J. P., Kahne, I., Jackson, R. L., and Marchesi, V. T.,** Major glycoprotein of the human erythrocyte membrane: evidence for an amphipathic molecular structure, *Arch. Biochem. Biophys.,* 155, 167, 1973.

27. **Nathans, J. and Hogness, D. S.,** Isolation and nucleotide sequence of the gene encoding human rhodopsin, *Proc. Natl. Acad. Sci. U.S.A.,* 81, 4851, 1984.

28. **Benedetti, E. L. and Emmelot, P.,** Studies on plasma membranes. IV. The ultrastructural localization and content of sialic acid in plasma membranes isolated from rat liver and hepatoma, *J. Cell Sci.,* 2, 499, 1967.

29. **Luft, J. H.,** Ruthenium red and violet. I. Chemistry, purification, methods of use for electron microscopy and mechanism of action, *Anat. Rec.,* 171, 347, 1971.

30. **Reid, B. L. and Charlson, A. J.,** Cytoplasmic and cell surface deoxyribonucleic acids with considerations of their origin, *Int. Rev. Cytol.,* 60, 27, 1979.

31. **Boyd, W. C.,** The proteins of immune reactions, in *The Proteins,* Vol. 2, Neurath, H. and Bailey, K., Eds., Academic Press, New York, 1954, 755.

32. **Kobiler, D., Beyer, E. C., and Barondes, S.,** Developmentally regulated lectins from chick muscle, brain and liver have similar chemical and immunological properties, *Dev. Biol.,* 64, 265, 1978.

33. **Nowak, T. P., Haywood, P. L., and Barondes, S. H.,** Developmentally regulated lectin in embryonic chick muscle and a myogenic cell line, *Biochem. Biophys. Res. Commun.,* 68, 650, 1976.

34. **Allen, A. K. and Neuberger, A.,** Purification and properties of the lectin from potato tubers, a hydroxyproline-containing glycoprotein, *Biochem. J.,* 135, 307, 1973.

35. **Nicolson, G. L.,** The interactions of lectins with animal cell surfaces, *Int. Rev. Cytol.,* 39, 89, 1974.

36. **Brown, J. C. and Hunt, R. C.,** Lectins, *Int. Rev. Cytol.,* 52, 277, 1978.

37. **Aub, J. C., Tieslau, C., and Lankester, A.,** Reactions of normal and tumour cell surfaces to enzymes. I. Wheat-germ lipase and associated mucopolysaccharides, *Proc. Natl. Acad. Sci. U.S.A.,* 50, 613, 1963.

38. **Burger, M. M. and Oldberg, R. A.,** Identification of a tumour-specific determinant on neoplastic cell surfaces, *Proc. Natl. Acad. Sci. U.S.A.,* 57, 359, 1967.

39. **Bhisey, A. N., Rao, S. G. A., Advani, S. H., and Ray, V.,** Agglutination of granulocytes from chronic myeloid leukaemia by Concanavalin A, *Acta Haematol.,* 63, 211, 1980.

40. **Nicolson, G. L., Poste, G., and Tae, H. Jl.,** The dynamics of cell membrane organization, *Cell Surface Rev.,* 3, 1977, chap. 1.

41. **Sainis, K. B. and Phondke, G. P.,** Confirmatory evidence for the existence of two types of receptors for Con A on mouse lymphocytes, *Immunology,* 40, 201, 1980.

42. **Sainis, K. B., Bhisey, A. N., Sundaram, K., and Phondke, G. P.,** The status of Concanavalin A receptors on the lymphocytes of leukaemic AKR mice: inhibition of redistribution by high concentrations of the lectin, *Cancer Biochem. Biophys.,* 6, 101, 1982.

43. **Joshi, N. N.,** *Biophysical Studies on Redistribution of Surface Receptors on Murine Lymphocytes,* Ph.D. thesis, University of Poona, India, 1984.

44. **Oppenheimer, S. B.,** Interactions of lectins with embryonic cell surface, *Curr. Top. Dev. Biol.,* 11, 1, 1977.

45. **Morré, D. J.,** The Golgi apparatus and membrane biogenesis, *Cell Surface Rev.,* 4, 1977, chap. 1.

46. **Lehninger, A. L.,** *Principles of Biochemistry,* Worth, New York, 1982.

47. **Horowitz, A. F.,** Manipulation of the lipid composition of cultured animal cells, *Cell Surface Rev.,* 3, 1977, chap. 5.

48. **Blobel, G. and Dobberstein, B.,** Transfer of proteins across membranes. I. Presence of proteolytically processed and unprocessed nascent immunoglobulin light chains on membrane-bound ribosomes of murine myeloma, *J. Cell Biol.,* 67, 835, 1975.

49. **Walter, P., Gilmore, R., and Blobel. G.,** Protein translocation across the endoplasmic reticulum, *Cell,* 38, 5, 1984.

50. **Meyer, D. I., Krause, F., and Dobberstein, B.,** Secretory protein translocation across membranes — the role of the "docking protein", *Nature (London),* 297, 647, 1982.

51. **Hanover, J. A. and Lennarz, W. J.,** Transmembrane assembly of membrane and secretory glycoproteins, *Arch. Biochem. Biophys.,* 211, 1, 1981.

52. **Roseman, S.,** The synthesis of complex carbohydrates by multiglycosyltransferase systems and their potential function in intercellular adhesion, *Chem. Phys. Lipids,* 5, 270, 1970.

53. **Pierce, M., Turley, E. A., and Roth, S.,** Cell surface glycosyltransferase activities, *Int. Rev. Cytol.,* 65, 1, 1980.

54. **Hirano, H., Parkhouse, B., Nicolson, G. L., Lennox, E. S., and Singer, S. J.,** Distribution of saccharide residues on membrane fragments from a myeloma cell homogenate: its implications for membrane biogenesis, *Proc. Natl. Acad. Sci. U.S.A.,* 69, 2945, 1972.

55. **Verkleij, A. J., Zwaal, R. F. A., Roelofson, B., Comfurius, P., Kastelijn, D., and Van Deenen, L. L. M.,** The asymmetric distribution of phospholipids in the human red cell membrane: a combined study using phospholipases and freeze etch electron microscopy, *Biochim. Biophys. Acta,* 323, 178, 1973.

56. **Bretscher, M. S.,** Membrane structure: some general principles: membranes are asymmetric lipid bilayers in which cytoplasmically synthesized proteins are dissolved, *Science,* 181, 622, 1973.

57. **Shinitzky, M. and Henkart, P.,** Fluidity of cell membranes — current concepts and trends, *Int. Rev. Cytol.,* 60, 121, 1979.

58. **Balian, R., Chabre, M., and Devaux, P. F.,** *Membranes and Intercellular Communication,* North-Holland, Amsterdam, 1982.

59. **Quinn, P. J.,** Fluidity of cell membranes and its regulation, *Prog. Biophys. Molec. Biol.,* 38, 1, 1981.

60. **Frye, L. D. and Edidin, M.,** The rapid intermixing of cell surface antigens after formation of mouse-human heterokaryons, *J. Cell Sci.,* 7, 319, 1970.

61. **Gall, W. E. and Edelman, G. M.,** Lateral diffusion of surface molecules in animal cells and tissues, *Science,* 213, 903, 1981.

62. **Nicolson, G. L.,** Transmembrane control of the receptors on normal and tumour cells. I. Cytoplasmic influence over cell surface components, *Biochim. Biophys. Acta,* 457, 57, 1976.

63. **Nicolson, G. L.,** Topographic display of cell surface components and their role in transmembrane signalling, *Curr. Top. Dev. Biol.,* 13, 305, 1979.

64. **Schmidt, M. F. G.,** Acylation of proteins — a new type of modification of membrane glycoproteins, *Trends Biochem. Sci.,* 7, 322, 1982.

65. **Smith, R. L. and Oldfield, E.,** Dynamic Structure of membranes by deuterium NMR, *Science,* 225, 280, 1984.

66. **Karnovsky, M. J., Kleinfeld, A. M., Hoover, R. L., and Klausner, R. D.,** The concept of lipid domains in membranes, *J. Cell Biol.,* 94, 1, 1982.

67. **Sherbet, G. V.,** personal communication, 1985.

68. **McNutt, N. S.,** Freeze-fracture techniques and applications to the structural analysis of the mammalian plasma membrane, in *Dynamic Aspects of Cell Surface Organization,* Cell Surface Reviews series, Vol. 3, Poste, G. and Nicholson, G. L., Eds., North-Holland, Amsterdam, 1977, chap. 2.

69. **Rash, J. E. and Hudson, C. S., Eds.,** *Freeze Fracture: Methods, Artifacts and Interpretations,* Raven Press, New York, 1979.

70. **Branton, D., Bullivant, S., Gilula, N. B., Karnovsky, M. J., Moor, H., Mühthaler, K., Northcote, D. H., Packer, L., Satir, B., Satir, P., Speth. V., Staehelin, L. A., Steere, R. L., and Weinstein, R. S.,** Freeze-etching nomenclature, *Science,* 190, 54, 1975.

71. **Singer, S. J. and Nicolson, G. L.,** The fluid mosaic model of the structure of cell membranes, *Science,* 175, 720, 1972.

72. **Perelson, A. S., Delisi, C., and Wiegel, F. W., Eds.,** *Cell Surface Dynamics: Concepts and Models,* Marcel Dekker, New York, 1984.

73. **Robertson, J. D.,** Membrane structure, *J. Cell Biol.,* 91, 189S, 1981.

74. **Good, N. E., Winget, G. D., Winter, W., Connolly, T. N., Izawa, S., and Singh, R. M. M.,** Hydrogen ion buffers for biological research, *Biochemistry,* 5, 467, 1966.

75. **Chaubal, K. A. and Lalwani, N. D.,** Electrical capacity of external cell surface: electrophoretic mobility analysis with polyanion treatment, *Indian J. Biochem. Biophys.,* 14, 285, 1977.

76. **Sherbet, G. V.,** *The Biophysical Characterization of the Cell Surface,* Academic Press, London, 1978.

77. **Hannig, K.,** Continuous free-flow electrophoresis as an analytical and preparative method in biology, *J. Chromatography,* 159, 183, 1978.

78. **Hannig, K., Wirth, H., Schindler, R., and Spiegel, K.,** An analytical version for rapid, quantitative determination of electrophoretic parameter, *Hoppe Seylers Z. Physiol. Chem.,* 358, 753, 1977.

79. **Mayhew, E. and Weiss, L.,** Ribonucleic acid at the periphery of different cell types and effect of growth rate on ionogenic groups in the periphery of cultured cells, *Exp. Cell Res.,* 50, 441, 1968.

80. **Pretlow, T. G. and Pretlow, T. P.,** Cell electrophoresis, *Int. Rev. Cytol.,* 61, 85, 1979.

81. **Shortman, K.,** The separation of lymphoid cells on the basis of physical parameters: separation of B- and T-cell subsets and characterization of B-cell differentiation stages, in *Methods of Cell Separation,* Catsimpoolas, N., Ed., Plenum Press, New York, 1977, 229.

82. **Iyengar, R., Mallman, D. S., and Sachs, G.,** Purification of distinct plasma membranes from canine renal medulla, *Am. J. Physiol.,* 234, F-247, 1978.

83. **Furchgott, R. F. and Ponder, E.,** Electrophoretic studies on human red blood cells, *J. Gen. Physiol.,* 24, 447, 1941.

84. **Vesterberg, O.,** Isoelectric fractionation analysis and characterization of ampholytes in natural pH gradients. V. Separation of myoglobins and studies on their electrochemical differences, *Acta Chem. Scand.,* 21, 206, 1967.

85. **Rao, K. V.,** Isoelectric focusing on HEPES-buffered pH gradients, *Indian J. Exp. Biol.,* 15, 552, 1977.

86. **Rao, K. V. and Duraiswami, S.,** Isoelectric focusing on pH gradients generated by zwitterionic buffer salts, *Indian J. Exp. Biol.,* 16, 1221, 1978.
87. **Boltz, R. C., Jr., Todd, P., Hammerstedt, R. H., Hymer, W. C., Thomson, C. J., and Docherty, J.,** Initial studies on separation of cells by density gradient isoelectric focusing, in *Cell Separation Methods,* Bloemendal, H., Ed., North-Holland, Amsterdam, 1977, 143.
88. **Arbuthnott, J. P. and Beeley, J. A., Eds.,** *Isoelectric Focusing,* Butterworths, London, 1975.
89. **Ave, K., Kawakami, I., and Sameshima, M.,** Studies on the heterogeneity of cell populations in amphibian presumptive epidermis with reference to primary induction, *Dev. Biol.,* 17, 617, 1968.
90. **Sherbet, G. V., Lakshmi, M. S., and Rao, K. V.,** Characterization of the ionogenic groups and estimation of the net negative charge on the surface of the cells using natural pH gradients, *Exp. Cell Res.,* 70, 113, 1972.
91. **Rao, K. V.,** Isoelectric focusing of cells on buffered pH gradients, *Indian J. Exp. Biol.,* 16, 724, 1978.
92. **Rao, K. V., Grover, R., and Mehta, A.,** Isoelectric focusing of cells using zwitterionic buffers, *Exp. Cell Biol.,* 47, 360, 1979.
93. **Rani, S., Dhar, S., Beohar, P. C., and Rao K. V.,** Isoelectric focusing: a new marker of differentiating T and B lymphocytes, *Indian J. Exp. Biol.,* 20, 172, 1982.
94. **Rao, K. V., Upadhyaya, K. C., Khan, M., Rafi, K. S., and Hasnain, M.,** unpublished data.
95. **Walter, H.,** Partition of cells in two-polymer aqueous phases: a method for separating cells and for obtaining information on their surface properties, in *Methods in Cell Biology,* Vol. 9, Prescott, D. M., Ed., Academic Press, New York, 1975, 25.
96. **Ballard, C. M., Roberts, M. H. W., and Dickinson, J. P.,** Cell surface charge correlation of partition with electrophoresis, *Biochim. Biophys. Acta,* 582, 102, 1979.
97. **Ballard, C. M., Dickinson, J. P., and Smith, J. J.,** Cell partition. A study of parameters affecting the partition phenomenon, *Biochim. Biophys. Acta,* 582, 89, 1979.
98. **Axen, R., Porath, J., and Ernback, S.,** Chemical coupling of peptides and proteins to polysaccharides by means of cyanogen halides, *Nature (London),* 214, 1302, 1967.
99. **Dvorak, D. J., Gipps, E., and Kidson, C.,** Isolation of specific neurones by affinity methods, *Nature (London),* 271, 564, 1978.
100. **Reisner, Y., Ravid, A., and Sharon, N.,** Use of soybean agglutinin for the separation of mouse B and T lymphocytes, *Biochem. Biophys. Res. Commun.,* 72, 1585, 1976.
101. **Bandyopadhyay, D.,** Separation of rat testis cells by a manually operated countercurrent distribution apparatus. I. Partition behaviour of cells in aqueous two-phase system, *Exp. Cell Res.,* 155, 557, 1984.
102. **Hamburger, V. and Hamilton, H. L.,** A series of normal stages in the development of the chick embryo, *J. Morphol.,* 88, 49, 1951.

Chapter 2

CELL ADHESION

I. INTRODUCTION

Multicellularity depends on mutual adhesiveness of cells. The cellular adhesive mechanisms have evidently evolved from the time multicellular animals came into existence. The complex histological architecture obviously depends on the adhesiveness of cells with other cells and with the intercellular substances. The mechanism by which cells adhere to each other and to the noncellular environment is therefore of a fundamental importance. Cleavage cells are held together by an effective adhesive mechanism. Embryogenesis involves cellular displacements in complex patterns.

At least a part of the mechanism controlling such dynamic behavior of cells must be the regulation of their adhesiveness. There are some cells that normally do not adhere to other structures. Erythrocytes are good examples of such free cells. Though normally "nonadhesive," these cells can adhere to each other (i.e., agglutinate) under certain circumstances such as the presence of incompatible blood group specific sera. Leukocytes are also free cells in the circulating blood, though they can also adhere to other tissue cells. Most of the higher invertebrates also have such free cells in their blood and coelomic fluids. These cells are normally nonadhesive. The sperm is another example of nonadhesive cells. A fully differentiated sperm does not adhere to any cell during its passage through the genital tract. The mammalian sperm penetrates the corona radiata, a cellular covering of the ovulated egg, without adhering to any cell or noncellular structure on its path. In fact, it adheres specifically only to one type of cell, viz., the unfertilized egg of the same species. These facts suggest *a priori* that the cell surface has some specific structural or molecular mechanisms of adhesion, i.e, the cells are not just "sticky" in a general way.

II. QUANTITATION OF ADHESIVENESS

A. Aggregation Kinetics

Quantitation of adhesiveness is one way of approach to understanding the molecular mechanisms involved in the mutual adhesion of cells. Attempts have been made to determine the adhesive strength of cell-glass (or cell-plastic) and cell-cell contacts using a variety of experimental techniques. Admittedly the cell-glass adhesion per se is not directly relevant to understanding the biological phenomenon; nevertheless, it has been studied extensively with the hope that it may reveal at least a part of the mechanism involved. The obvious advantage of determining cell-glass adhesion is the ease of experimental manipulation. Adhesion of cells to glass or plastic is enhanced by some serum components. Cells are also known to secrete some macromolecules at the points of contact with the inert surface. In view of this, even cell-glass adhesion is not to be considered as direct attachment to glass.

The percentage of cells adhering to glass in a given time has been used as a measure of cell adhesiveness by many workers.[1-3] Essentially the method consists of placing a cell suspension in glass containers and decanting it after different intervals of time. The percentage of cells adhered to glass is then determined. An obvious defect inherent in this methodology is that it does not take into account the time taken by the cells to reach the glass surface during sedimentation before the establishment of adhesive contacts could begin. Since not all cells start establishing cell-substratum contacts simultaneously, the quantitative determination includes an important error. In general, larger cells sink to the bottom of the container more rapidly than smaller ones and would therefore tend to appear more strongly

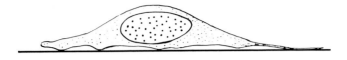

FIGURE 1. Diagram to show the spreading and adhesion of a cell on a glass surface. Note that the cell surface does not adhere to the glass all over its lower side.

adhesive. Any variation of this experimental technique is subject to criticism on another score. It has been shown that when the cell spreads and adheres to glass, the actual points of contact between the plasma membrane and the substratum are separated by regions of no contact between the two[4] (see Figure 1). In tissue cells, however, the situation differs from this in two respects. First, the area of contact between adjacent cells in compact cellular tissues is much larger and is not discontinuous. Second, the overall shape of the cells after adhering to glass is often different from their native shapes in cell clusters in vivo. All results obtained from studies on cell-glass adhesion will have to be interpreted bearing these points in mind. Work on the adhesion of cells to glass, plastic, and other artificial substrata has been reviewed by Grinnel[5] and Curtis et al.[6]

Determination of the adhesive strength in cell-cell contacts is undoubtedly more relevant biologically. There are, however, many difficulties in designing suitable experimental techniques. Most of the methods currently in practice are those that use previously disaggregated cells in suspension. On stirring such a suspension, cells collide against each other, and continuously growing aggregates are formed. A rough estimate of cell adhesiveness can be obtained from the average diameter of the aggregates formed at a given time. Since the aggregate size is widely variable and would depend on several uncontrolled factors, the percentage of single cells remaining behind in the stirred suspension may be used as a more accurate estimate of adhesiveness. Extensive experimental work has been done using this method.[7,8]

The methods described above yield reproducible data and are satisfactory for most work. A more accurate estimate of adhesiveness can, however, be obtained only by a closer analysis of the events occurring in the stirred cell suspension. In any technique of reaggregation of cells in shaking suspensions, the net number of adhesions between cells will depend on the total number of collisions that occur during a given time multiplied by a certain factor expressing the probability that any collision will result in a stable adhesion.[9] The total number of collisions is determined by the fraction of the volume of the agitating suspension occupied by the cells, the speed of agitation, and the viscosity of the medium. When these extrinsic variables are controlled, the aggregation process depends only on the intrinsic cell property, viz., the adhesiveness.[10] The size of aggregates formed is influenced by the shear force exerted on them, tending to tear apart the cells from the aggregates. Therefore the aggregate is protected by media of higher viscosity which, on the other hand, tend to decrease the frequency of collisions. If cells are shaken in media of high viscosity, their collision rate will be low, and therefore the rate of reaggregation will be low. However, larger aggregates will survive against shear in the more viscous media. Such a situation will therefore lead to a wrong impression about cellular adhesiveness if aggregate size is the measure.[9]

An improved method of assessing the process of cell reaggregation is to monitor the early events of the formation of aggregates. When a cell suspension is stirred, single cells collide against each other, and some of the collisions result in adhesion. The two-cell aggregates collide with other single cells and thus "grow" into larger aggregates simultaneously with increase in their number due to continued collision of single cells. If the adhesiveness of cells does not alter during the process of stirring, the kinetics of reaggregation should proceed in a predictable manner. The early work of Curtis and Greaves[11] using this approach was further improved in later studies.[9] A somewhat elaborate instrument was designed to eliminate inaccuracies involved in reciprocal or gyratory shaking. The "cuvette viscometer" ensures a constant and measureable laminar flow of the cell suspension held in a space between two concentric cylinders, one of which is rotating against the other.[9] In such a system, when experimental conditions like the composition of the medium and shear rate (due to the laminar flow of the cell suspension against a rotating wall) are constant, the total number of particles (cells and their aggregates) at the beginning of reaggregation, $N_{\infty 0}$ and after an interval of time t, $(N_{\infty t})$ are related as follows

$$\ln \frac{N_{\infty t}}{N_{\infty 0}} = \frac{-4\,\phi\,\alpha\,t}{\pi} \tag{1}$$

where ϕ is the volume fraction occupied by the particles and α is the "stability ratio". The probability that a collision between two cells results in adhesion is expressed by α, which can be given as

$$\alpha = -\ln \frac{N_{\infty t}}{N_{\infty 0}} \times \frac{\pi}{4\,\phi\,t} \tag{2}$$

Rao et al.[12] (see also George and Rao)[13] have generally followed the theoretical arguments of Curtis[9] and analyzed the early process of cell aggregation, treating it as analogus to simple chemical kinetics. When stirring starts, the major event is the formation of two-cell aggregates, their depletion in the formation of larger aggregates being negligible (Figure 2). So a reaggregation rate constant (k) during this initial phase can be calculated using the following equation of chemical kinetics:

$$k = \frac{1}{t} \times \frac{x}{a(a - x)} \tag{3}$$

where a is the density (number/volume) of single cells at the beginning of aggregation and $(a - x)$ at time t. The values of k can be normalized for an arbitrarily chosen constant value of the fraction of the total volume of cell suspension occupied by the cells. This permits direct comparison of the adhesiveness (in fact, k) of different tissue cells or similar cells before and after any experimental alteration.

Monitoring cell aggregation can be done by various methods. The simplest and most direct method is microscopic examination of samples drawn at intervals from the stirring suspension. Particle counters such as the Coulter counter are more accurate than hemocytometers. Besides, signals from such electronic instruments can be processed by computers directly. Population densities of cells and aggregates can be measured with hemocytometers using image analyzing computers.[14] Thomas and Steinberg[15] designed an automatic monitoring device consisting of as many as 12 cuvettes containing cell suspensions. These cuvettes rotate in sequence through a light beam. The decrease in small-angle light scattering, which accompanies cell aggregation, is monitored by a photomultiplier whose pulses are eventually recorded.

B. Adhesive Specificity and Its Measurement

Regardless of the sophistication of instrumentation resulting in the ease of operation and accuracy, the basic principle of all the methods described above is the same, viz., to determine the collision efficiency. In almost all studies, the cells used are homotypic (i.e., derived from the same embryonic or adult organ). In many instances, however, the information

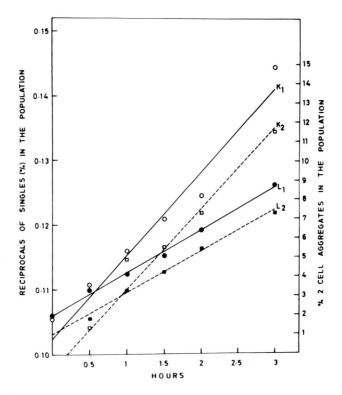

FIGURE 2.   Changes during the reaggregation of cells in a stirred suspension. A progressive depletion of cells accompanied by a parallel increase in the number of two-cell aggregates is observed. Reciprocals of single cells (%): ○, chick embryonic kidney cells; ●, chick embryonic liver cells; the corresponding two-cell aggregates (%) are □, kidney, and ■, liver. The lines K_1, L_1, and K_2, L_2 are linear regressions calculated for the corresponding data. (From George, J. V. and Rao, K. V., *J. Cell Physiol.*, 85, 547, 1975. With permission.)

needed is the adhesive strength in homotypic as well as heterotypic adhesions. Curtis[9,16] has shown that specificity of cell adhesion can be determined using the method of reaggregation in stirred suspensions. For this purpose, the collision efficiency is determined for two cell types separately and then for mixtures of various proportion of the two cells' types. If there is no specificity of adhesion, the collision efficiency for any proportion of the two cell types will be the average of the efficiencies for both types (measured separately and weighted for the proportion of each cell type in the mixture). If, however, there is complete specificity of adhesion, such that the two cell types do not adhere mutually, the measured collision efficiencies for a mixture of the two types will be $n_1^2 E_1 + n_2^2 E_2$, where E_1 and E_2 are, respectively, the collision efficiencies of the two cell types, and n_1 and n_2 are the numbers of cells taken in the mixture. If specificity is incomplete, the plots of collision efficiency against proportion of cell types within the mixture will lie between those for complete specificity and complete lack of it. Curtis[9,16] has described the statistical methods for the analysis of data obtained in such experiments.

It is not certain if the property measured by the various reaggregation methods described above is directly related to the strength of adhesion or specificity between cells. Obviously, cells do not collide against one another in the formation of tissues. It is, however, very likely that the readiness with which cells aggregate is an expression as well as a measure of their adhesive property.

Another avenue of approach to understanding the mechanism of cell adhesion and its quantitative determination has been followed by some workers, especially to understand specific cell adhesions. Cells seem to adhere preferentially to certain cells and show varying degrees of specificity. Though a number of different types of cells can be held together by experimentally bringing them close, a more specific adhesion seems to occur in tissues. In order to determine the specificity of adhesion and its quantitation, a novel experimental method was developed by Roth and Weston.[17] In this method, a cell aggregate of a certain size is released into a suspension of single cells, the later being recognizable from the former either by radioactive label or any other marker. When the cell suspension is stirred, the single cells are attached to the collecting aggregate. The aggregate is then sectioned and examined microscopically. If the single cells are radioactively labeled, autoradiography is necessary. The use of carrier-free phosphoric acid ^{32}P to label the cells and estimation of the radioactivity picked up by the collecting aggregate by scintillation counting makes the experiments rapid.[18] If the cells have any other marker, the appropriate cytological technique has to be used. The number of cells picked up by the collecting aggregate is a measure of the adhesiveness of the disaggregated cell type to those of the collecting aggregate. It has been shown that the collecting aggregate picks up more homotypic cells than heterotypic ones, showing specificity in adhesiveness.

Recently Koziol et al.[19] have developed a method of quantitative determination of selective cell adhesion using a different type of assay. In their method, cells of one type are fluorescently labeled green with fluorescein isothiocyanate and those of another type red with tetramethyl rhodamine isothiocyanate. When a suspension containing equal numbers of such green and red fluorescing cells is stirred, aggregates consisting of both cell types are formed. The "purity" of the aggregates is calculated as a purity index (PI). If all aggregates have cells of either only red or green fluorescence, the purity index is 1.0. If the red and green cells are in equal numbers the PI is 0.5. Values between 0.5 and 1.0 express greater degree of specificity in adhesion. The purity index is defined as

$$PI = N^{-1} \sum_{i=1}^{N} max(R_i \ G_i)/(R_i + G_i) \tag{4}$$

where R_i and G_i are, respectively, the numbers of red and green cells in the ith aggregate, and N is the number of aggregates observed (sample size). The PI does not take into account the aggregate size. No doubt an aggregate consisting of 20 red cells and none of green ones would indicate greater specificity for homotypic adhesion than one with two red cells and none of green. There are, however, statistical methods which can analyze such data and yield an "odds ratio" that permits comparison of adhesive specificities of varying degrees. Reference may be made to the paper of Koziol et al.[19] for full details. Based on more or less similar theoretical consideration, Sieber and Roseman[20] have described a method of quantitative estimation of adhesive specificity.

We have dealt with the various methods of quantitative determination of cellular adhesiveness rather at length to emphasize the point that any explanation of developmental or morphogenetic changes depending on differences in adhesive strength of cells or the spec-ificities of mutual adhesion should be based on reliable and comparable data. Extant literature is replete with reports on quantitative determination of cellular adhesiveness. Most of them are at best qualitative assays. Besides, data obtained by one group of workers cannot be compared with those from others because of numerous differences in the experimental techniques and sometimes even in the definition of adhesiveness. It is quite probable that different cell properties are assayed by the different methods, and they are all related to the in vivo adhesiveness of cells only indirectly. Future attempts at measuring the adhesiveness of cells should take into account an important outcome of recent research. As we shall see later in this chapter, there are more than one distinct molecular mechanism of adhesion even in a single cell type. Unless we have a method to determine the strength of the individual adhesive mechanisms, we cannot get a full idea of how cells make contact choices. Though the recent methods such as those of Koziol et al.[19] and Sieber and Roseman[20] have the necessary precision, they can yield relevant data only if the molecular mechanisms of adhesion can be dissected further and examined individually.

III. EXPERIMENTAL STUDIES ON CELL ADHESION

Experimental attempts at revealing the molecular mechanism of cell adhesion have chiefly relied on the measurements of cell adhesiveness, by the various assay methods outlined above. Information obtained from cell-glass adhesions cannot be dismissed as irrelevant, since conclusions from such studies often also accord with those based on cell-cell adhesions. A very large volume of literature on the subject of cell-cell and cell-inert substrate adhesion is available. The brief account given here is based on investigations that have been generally confirmed by several independent groups of investigators.

Cells do not adhere readily to inert objects such as plastics. Wettability is an essential quality of the substratum.[21] Many types of cells adhere only to plastic or glass coated with some protein such as collagen or gelatin. From these observations, it follows that cells do not adhere to hydrophobic surfaces. It is also clear that the nature of the substratum has an important role. In view of this, it seems more appropriate to determine cell adhesiveness on glass surfaces coated previously with naturally produced extracellular matrix material. It has been shown that endothelial cells grown on glass/plastic substrata, and removed subsequently, leave behind some complex extracellular matrix that seems to be similar to the basement membrane in composition and supramolecular arrangement. Culture dishes coated with such naturally produced extracellular matrix promote better attachment, proliferation, and differentiation of cells. Even the serum requirement is lowered. It has been claimed that the physiological responses of such cells are also more approximating to natural ones.[22,23]

A. The Role of Temperature in Cell Adhesion

Considerable work has been done to examine in general the role played by temperature in the control of cell reaggregation. Many workers have independently demonstrated that lower temperatures are inhibitory to cell reaggregation.[2, 24-26] The inhibitory effect could be due to two different reasons: (1) low temperature may diminish physiological activity, especially energy metabolism, and thus lead to decreased adhesiveness; (2) the surface properties of the cells may alter at low temperatures, and this may be the cause of decreased adhesiveness. In connection with the first of these explanations, it is necessary to examine if cellular energy production is indeed necessary for establishment of adhesive contacts. If later events of cell aggregation such as increase of the aggregates' size are taken as measures of adhesiveness, energy production seems to be necessary.[7,8] On the other hand, the initial adhesion due to collisions seems to occur without energy requirements.[12,27] The question seems to be still unsettled,[28] especially because the experimental methods used by different workers are not comparable.

The second explanation, viz., alteration of the surface properties, seems plausible, especially in view of the more recent findings on the biophysical changes undergone by the plasma membrane components. George and Rao[27] demonstrated that chick embryonic kidney and liver cells reaggregate very slowly at 4°C. Up to 24°C, there is an increase in the reaggregation rate with rise of temperature. The range of temperature that allows the highest rate of reaggregation is 24 to 37°C. At temperatures higher than physiological, the rate of reaggregation is diminished to about 50% of the maximum. Cells exposed to 40°C do not regain normal rate of reaggregation even if stirred at 37°C (Figure 3). Since reaggregation of these cells was not inhibited by azide, 2-4-dinitrophenol, and rotenone, it was interpreted that the decreased reaggregation at low temperature was due to some change in the cell surface properties. Low temperature depolymerizes microtubules, which renders the cell

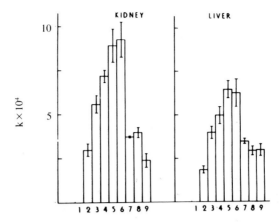

FIGURE 3. Influence of temperature on the rate of reaggregation of chick embryonic kidney and liver cells. *Ordinate:* the reaggregation rate constant (*k*) per hour, was calculated as in Equation 3. The vertical bars on the abscissa represent reaggregation at various temperatures: 1, 4°C; 2, 8°C; 3, 16°C; 4, 20°C; 5, 24°C; 6, 37°C; 7, 40°C; 8, cells exposed to heat shock at 40°C for 2 hr and then reaggregated at 37°C; 9, 43°C. Each value is a mean of at least five replicates, ±SE. (From George, J. V. and Rao, K. V., *Indian J. Exp. Biol.*, 13, 470, 1975. With permission.)

more spherical. Besides, the lipid layers of the membrane are more solid at low temperature and thus tend to decrease deformability. Once the adhesive contact is established, it is not disrupted by low temperature.

It is well-known that cultured cells adhering to glass can be exposed to a low temperature without resulting in their detachment. Similarly, cells already aggregated into clusters do not fall apart at low temperatures. The irreversible change in the reaggregation rate of cells exposed to high temperature (Figure 3) is, in all probability, due to extensive rearrangements in the lipids and other components of the plasma membrane. Thus the effect of temperature is not on the maintenance of adhesive bonds between cells; it influences single cells in suspension, and the decreased reaggregation rate is due to failure of the formation of proper contacts between cells with modified surface properties.

B. The Role of Divalent Cations in Cell Adhesion

The divalent cations Ca^{2+} and Mg^{2+} are known to be important for cell adhesion. Earlier researchers concluded this from several observations, including the ease of cell disaggregation in calcium-magnesium-free (CMF) media and lack of adhesiveness in the presence of chelating agents such as ethylenediaminetetraacetic acid (EDTA). It was suggested that calcium could form a bridge between covalent binding sites of two cells (Figure 4).[29,30] Curtis[31] suggested that calcium can bind on negatively charged groups such as COO^- on the cell surface and thus decrease the electrostatic repulsive forces. As we shall see in a later section, the cells can adhere to specific molecules constituting the extracellular matrix. Binding with these molecules may involve a definite role for Ca^{2+}. There is also evidence that cell surface receptors mediating in cell adhesion may be protected against proteolysis by Ca^{2+}.[32]

Recent work has shown that the role of the divalent cations is not simple. Takeichi et al.[33] (see also Urushihara et al.[34]) have demonstrated that experimentally separable Ca-dependent and Ca-independent mechanisms exist in cultured Chinese hamster ovary and chick neural retina cells. Similar observations have been reported by Lilien and associates[35,36] and Steinberg and associates.[37-39] These workers' observations are largely in agreement with each other. Neural retina cells of chick embryos disaggregated in the presence of EDTA as well as trypsin are nonadhesive. However, if disaggregated in the presence of trypsin or EDTA alone, they can still reaggregate. Even cells disaggregated in the presence of EDTA and low trypsin can reaggregate (see Figure 5).

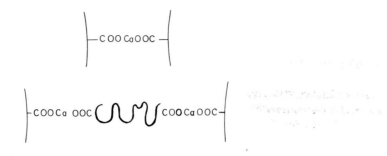

FIGURE 4.   The calcium bridge hypothesis of cell adhesion. Ca^{2+} ions bind COOH groups exposed on the cell surface and may bring about adhesion of cells. Two alternatives are shown: direct bridging (above) and adhesion through a link molecule, represented by the thick wavy line (below).

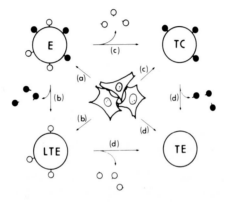

FIGURE 5.   Schematic representation of "adhesion sites" in V79 cells. Open circles represent Ca^{2+}-independent adhesion sites (CIDS) and closed circles, Ca^{2+}-dependent sites (CDS). From cultured cells shown at the center, four kinds of dispersed cells with different conditions on the cell surface can be obtained. E, cells with both sites intact; LTE, cells with CIDS only; TC, cells with CDS only; TE, cells with neither binding sites. Arrows represent treatment with (a) EDTA, (b) light trypsin-EDTA, (c) trypsin-Ca^{2+} and (d) trypsin-EDTA. (From Urushihara, H., Okazaki, H. S., and Takeichi, M., *Dev. Biol.*, 70, 206, 1979. With permission.)

From these observations, it has been concluded that there are two distinct adhesive mechanisms, one of them being calcium independent and the other dependent. The precise molecular mechanism by which the two operate is still somewhat speculative. The explanations given by all the three independent groups of workers agreed generally with the following account. It is proposed that there are two kinds of cell surface molecular sites constituting the calcium dependent and independent mechanisms (see Figure 5). When the cells are treated with EDTA, they retain both the adhesive mechanisms. If such cells are now treated with trypsin, they retain only the Ca-dependent mechanism. On the other hand, if they are treated with a low concentration of trypsin in the presence of EDTA, the calcium-dependent mechanism is lost, but the calcium-independent one is retained. The cells that retain one or the other mechanism during disaggregation are able to reaggregate, whereas those lacking both are nonadhesive. How the treatments remove selectively one or the other mechanism is not yet understood fully.

Are the Ca-dependent and independent mechanisms indeed distinct and coexistent on the cell surface? If they coexist, are the two functionally dependent on one another? These questions have been answered by the following experimental results.[38,39] Monovalent Fab molecules were prepared from antibodies raised against cells bearing only one or the other mechanism. Reaggregation of cells retaining one of the putative mechanisms was inhibited by the addition of Fab against the respective mechanism, but not the other. Fluorescently labeled Fab molecules can specifically bind the respective mechanisms. Another observation is that the cells with one mechanism cannot reaggregate with cells having only the other. Thus the mechanisms seem to be functionally specific and independent. In other words, the two mechanisms do not act as lock and key components of a single system of adhesive mechanism.

These interesting observations raise the question of the functional significance of the dual adhesive mechanism during the morphogenetic rearrangements of cells. It is tempting to speculate that modulation of one or both the mechanisms could bring about morphogenetic changes of cell association in embryogenesis. At present, there is no evidence to suggest that each of the mechanisms is itself not uniform in case of all cell types, though this possibility cannot be ruled out. More work is needed to understand the significance of these findings.[40]

C. Cell Surface Molecules and Adhesion

Adhesion is obviously mediated by the cell surface molecules that have to react in some way with the surface to which they adhere. Attempts have been made to show that proteins or glycoproteins of the cell surface are an essential component of the adhesion mechanism. It is generally known that trypsinization, which is often used as a method of cell disaggregation, renders the cells inadhesive for some time. A "lag" of 30 min to 4 hr in reaggregation of cells due to trypsinization has been reported by different workers. It is known that even a very brief exposure of cells to trypsin can affect the plasma membrane glycoproteins. Other proteases are equally harsh if not more. There is evidence to show that fresh protein synthesis is essential for the recovery from the "lag" phase.[41] The use of other proteolytic enzymes also shows the lag in reaggregation.

The exact role of the proteins in establishing an adhesive contact with the substratum or another cell is, however, not clear. Many attempts have been made to understand this, especially by modifying the surface proteins. A large number of workers have shown that alkylation of the free -SH groups brings about a decrease in adhesion. Ideally, a specific reagent binding permanently and covalently on the cell surface protein $-$SH residues should be chosen for such studies. Inhibition of cell adhesion by alkylating agents such as mercuric chloride and *N*-ethyl maleimide[3,41] is often difficult to interpret, since they can penetrate into the cytoplasm and act on components other than the plasma membrane. The possible effect of such agents on the respiratory metabolism of the cells does not seem to be responsible for the inhibition of reaggregation by the $-$SH binding agents. This is concluded from the fact that other specific inhibitors of respiratory activity do not affect the initial reaggregation of cells. However, other possible effects of the $-$SH binding agents, which can diffuse into the cytoplasm, cannot be ruled out. Any alteration of cell adhesion could therefore be a secondary effect. It is known that $-$SH binding agents can inhibit tubulin polymerization resulting in cell rigidity. An interesting $-$SH reagent is carboxypyridine disulfide (CPDS), which at low concentrations and short duration of treatment does not penetrate the plasma membrane.[42,43] The reaction of the CPDS molecule on the surface $-$SH residues is shown in Figure 6. As shown in the figure, the action of CPDS involves alkylation of the $-$SH group and also the introduction of a $-$COOH group. George and Rao[13] showed that CPDS reacts with $-$SH groups (presumably at the surface) and renders the cells inadhesive. This action could be readily reversed by a subsequent treatment with certain thiols (Figure 7).

There are other $-$SH blocking agents such as ω-chloroacetophenone and *N*-ethyl mal-

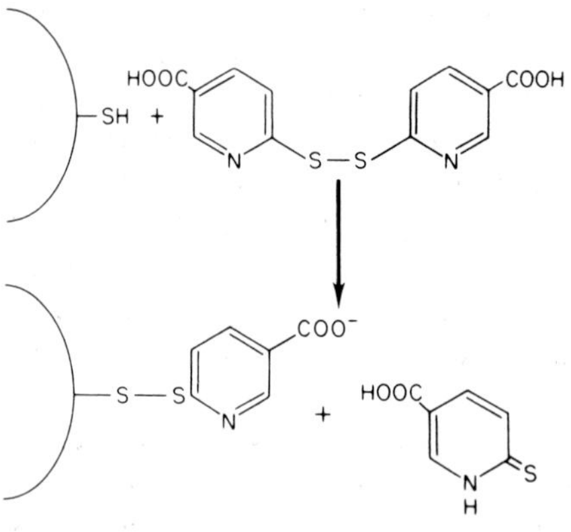

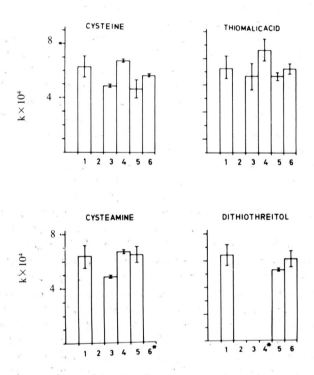

FIGURE 6. Binding of carboxypyridine disulphide (CPDS) to the cell surface. The disulphide bond of the CPDS molecule splits up and one half forms a disulphide bridge with the cell surface thiol group resulting in the introduction of one negative charge from each thiol group reacted. (From Sherbet, *The Biophysical Characterization of the Cell Surface*, Academic Press, Orlando. With permission.)

FIGURE 7. Effect of binding thiol groups of chick embryonic liver cell by carboxypyridine disulphide (CPDS) and its regeneration by thiols. *Ordinate;* the reaggregation rate constant *(k)* per hour, calculated as in Equation 3. *Abscissa:* the vertical bars 1 to 6 represent different experimental protocols: 1, untreated cells; 2, cells treated with $2 \times 10^{-4} M$ CPDS for 5 min, washed and reaggregated; 3, cells treated as in 2, washed and resuspended in the reaggregation medium containing $2 \times 10^{-4} M$ thiols indicated; 4, cells treated as in 2, exposed to the thiols $(2 \times 10^{-4} M)$ for 15 min, washed and resuspended in the reaggregation medium; 5, cells suspended in the reaggregation medium containing the thiols indicated; 6, cells exposed to the thiols indicated for 15 min, washed and reaggregated. * — These experiments were not done; each bar is a mean of at least five replicates ±SE. (From George, J. V. and Rao, K. V., *J. Cell Physiol.*, 85, 547, 1975. With permission.)

eimide whose action on cell adhesion is similar to that of CPDS, but irreversible by the thiols added subsequently.[44] The observation that covalent binding of $-SH$ groups inhibits cell adhesion cannot, however, be interpreted to mean that the adhesive mechanism is solely dependent on them. Studies such as those mentioned above merely indicate that a slight modification of the cell surface leads to an interference with the mechanism of adhesion. They do not, however, identify any particular molecular species constituting the adhesive mechanism.

D. The Extracellular Matrix and Cell Adhesion

Any discussion on cell adhesion in vivo has to take into account the material found in the intercellular spaces constituting a matrix. The old idea that the extracellular matrix is secreted by the cells as a mechanical scaffolding material is no longer valid. Numerous studies have indicated clearly that there is a constant interaction between the cells and the molecules constituting the matrix in which the former respond not only by adhering but also by changing their internal activities. Thus modifications in the organization of the cytoskeletal structures resulting in changes in the cells' shape, and control of synthetic activities resulting in growth and differentiation are direct consequences of such interactions between the cells and their extracellular material. Studies on its chemical composition and the nature of interactions relevant to morphogenesis are now major attempts in biological research.

Four major classes of macromolecules are recognized in the extracellular matrix: collagen, elastin, proteoglycans, and glycoproteins. In this section, we give a brief account of their chemistry and indicate how they are linked with each other and with the cells.

1. Collagen

Collagen is the major component of the intercellular material in the vertebrate body. Collagenous proteins occur in all animals except Protozoa. A unique feature of collagen is its amino acid composition. Glycine forms very close to one third the total amino acid residues; proline and hydroxyproline form another 25%. Its long chain polypeptides consisting of about a thousand amino acid residues exist as a rod-like triple helix.[45] The helical chains are strengthened by interstrand hydrogen bonds. Native collagen solutions are viscous due to the molecular dimensions, viz., ≈ 3000 Å long and ≈ 15Å diameter. Both the N- and C-terminal ends of the polypeptides have short (≈ 20 amino acid residues) nonhelical stretches (Figure 8). These are the regions that form intermolecular and intramolecular cross-links in native collagen. Each polypeptide has a sequence of amino acids in the pattern of gly-X-Y where X and Y are any other amino acids, though frequently X is proline and Y is hydroxyproline. Other amino acids may occupy the positions X and Y in the triplet less frequently. Another striking feature of the polypeptides is the posttranslational hydroxylation of their proline and lysine residues. Lysines are also glycosylated. The lysine residues of the nonhelical region are hydroxylated and, by further enzymic modification, can form cross-links with other polypeptides.

There are at least five different types of collagen in the human body. They are named collagen I to V (see Table 1). Collagens I to III are components of different connective tissues. Collagen IV is associated with basement membranes of epithelia. Type V is associated with the synovial membrane and some other structures. It should be noted, however, that collagen types probably do not have a unique distribution, though quantitative differences in proportions of the different types in local anatomic sites exist. It is obviously these differences that are important for tissue function. Modulated synthesis of the types is presumably the basis of tissue differentiation. Collagen is secreted by connective tissue cells and deposited as highly organized fibrils. The fact that collagen types have characteristic tissue distribution and that the process of deposition in the intercellular space involves several enzyme-mediated steps indicates that wide structural variations are possible by modulating

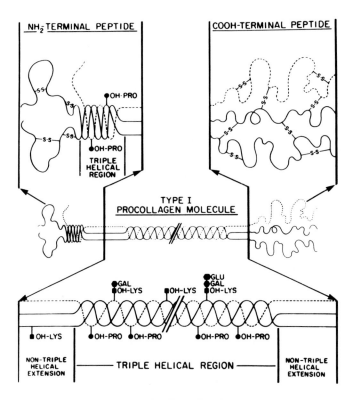

FIGURE 8. Diagram of the major characteristics of the Type I collagen molecule and its procollagen form. OH-PRO, hydroxyproline; OH-LYS, hydroxylysine; $-S-S-$, disulfide bonds; GAL, galactose. (From Linsenmayer, T. F., *Cell Biology of Extracellular Matrix,* Hay, E. D., Ed., Plenum Press, 1981, chap. 1. With permission.)

Table 1
TYPES OF COLLAGEN AND THEIR TISSUE DISTRIBUTION

Collagen type	Chains	Typical tissue source
I	α1, α2	Bone, tendon, dermis
II	α1	Cartilage, intervertebral disc
III	α1	Hyaline matrix of cartilage, dermis
IV	?	Basement membrane of epithelia
V	α1, α2 (?)	Mouse blastocyst and teratocarcinoma cells

the different steps. Besides, the collagen molecules and their aggregates interact with other components of the intercellular material and thereby provide for a very large variation in the final product, viz., the extracellular matrix.[46] Cell surface associated proteins are known to have collagen binding specificity,[47] which may be of significance in vivo. Collagen research has assumed importance in diverse biological areas and the huge literature that has appeared in the past two decades is summarized in the following books and reviews: Ramachandran and Reddi;[48] Bornstein and Sage;[49] Linsenmayer;[50] and Piez and Reddi.[51]

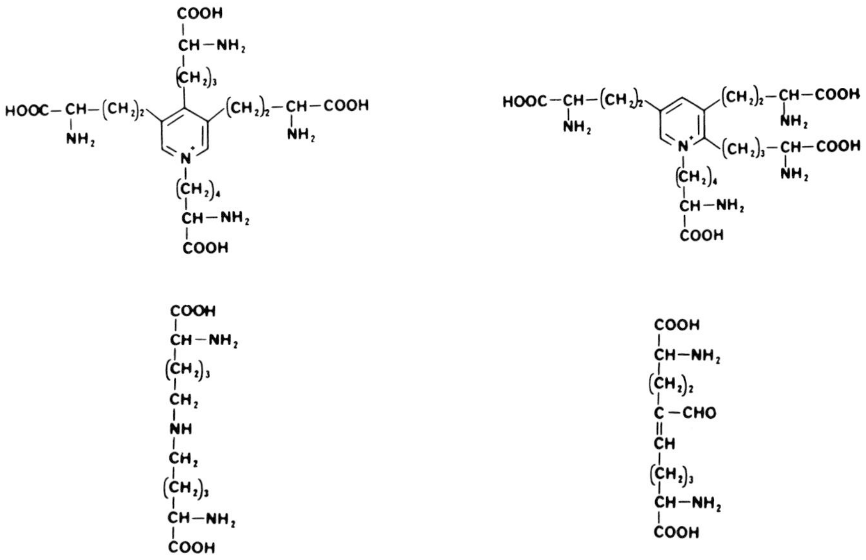

FIGURE 9.   Unusual amino acids found in elastin. Upper, left: desmosine; upper, right: isodesmosine; lower, left: lysinonorleucine; lower, right: aldol condensation product of two allysine residues.

2. Elastin

Elastin is an insoluble protein occurring in the extracellular matrix. It is familiar as elastic noncollagenous protein fibers in the dermis, trachea, and many other tissues. Extraction and purification of elastin is not easy and involves very harsh treatments like boiling in dilute alkalis or autoclaving. Extant information on the chemistry of elastin is therefore unsatisfactory. The amino acid composition of elastin is unique: one third of all residues are glycine, one ninth are proline, and the other major components are alanine, valine, leucine, isoleucine, and phenylalanine. Well over 90% of all amino acid residues of elastin are hydrophobic. This is in accord with the fact that elastin is a highly insoluble protein.

Another important feature of the amino acid composition is the occurrence of several unusual lysine-derived amino acids[52,53] (see Figure 9), which form interchain cross-links by covalent bonding. The identity of the cells which produce elastin is not clear, though aortic smooth muscle cells have been shown to be able to elaborate this substance in cultures. Possibly other cell types are also responsible for the synthesis of elastin. It is known that a soluble precursor molecule (tropoelastin) is first secreted. It is then organized into the cross-linked insoluble elastin. There is evidence that the formation of insoluble elastin depends on some, not yet clearly understood, role of other macromolecules of the extracellular matrix. For a recent review on the various aspects of elastin, see Franzblau and Faris.[53] Besides its obvious function by virtue of its mechanical property, elastin may influence the motile behavior of cells, as suggested by the work of Senior et al.[54]

3. Proteoglycans

Proteoglycans are complex macromolecules of the extracellular matrix found in almost all tissues, especially in the connective tissues. They consist of a core protein to which are attached covalently bound glycosaminoglycans. There are different core proteins and also a vast variety of glycosaminoglycans. Thus proteoglycans constitute a vastly variable class of macromolecules serving diverse structural and organizational roles in tissues. Cartilage is a very good source of proteoglycans.

The structure of these proteoglycans is very complex (Figure 10). These are "enormous" molecules with a backbone protein of $\approx 250,000$ mol wt on which are bound through *O-*

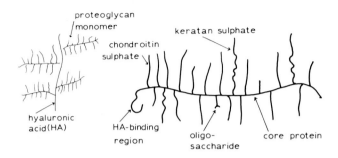

FIGURE 10.   Diagram representing the structure of complex proteoglycans. On the left is shown a part of the proteoglycan complex consisting
of four monomers bound to a molecule of hyaluronic acid. On the right
is shown a monomer consisting of the core protein and other molecules
bound on it.

linkage about 80 chondroitin sulfate chains and keratan sulfate and "mucin-type" oligosaccharides ($\approx$100 per core protein) besides some N-linked oligosaccharides. One end of
the core protein that is relatively free of the glycosaminoglycans is linked to hyaluronic
acid. The synthesis of these substances is a complex process. Elucidating the orchestration
of the pathways and their control in proteoglycan synthesis is a formidable problem of
biochemistry. Like any other secreted protein, the core protein is synthesized on the ribosomes, and after its transport across the membrane, it is glycosylated in the Golgi. Secretion
is effected by exocytosis. The secretory vesicles containing the glycosylated proteins reach
the plasma membrane and fuse with it, releasing their contents. Proteoglycan aggregate
formation (i.e., the association of glycosylated core proteins with hyaluronic acid) occurs
extracellularly, probably through an interaction with collagen.

From the functional viewpoint, a peculiar feature of the proteoglycans has to be mentioned.
In solution, these macromolecules have an extended structure occupying a very large hy-
drodynamic volume relative to their molecular weight. They also have a large number of
negative charges. They can hold a very large volume of water and ions. These molecules
are reversibly compressible, a property critical for cartilage. The proteoglycans occupy large
spaces that appear "empty" in tissue sections prepared for routine histological examination.
For more information on proteoglycans, see Hascall and Hascall[55] and Hardingham et al.[56]

4. Glycoproteins

Besides the collagen, elastin, and proteoglycans, the extracellular matrix contains several
important glycoproteins. They have interesting properties that include roles in cell adhesion
as well as interactions with the other components of the extracellular matrix. Fibronectin,
laminin, and chondronectin are the best known among them. It is, however, likely that more
such proteins or glycoproteins and their roles will be identified in the near future.

a. Fibronectin

Fibronectin is a large extracellular glycoprotein. It binds to a variety of molecules in the
extracellular milieu. These include fibrinogen/fibrin, collagen, heparin and heparan sulfate,
transglutaminase substrates, and hyaluronic acid. It also binds on cell surfaces (especially
fibroblasts), actin, DNA, and bacteria (*Staphylococcus aureus*). Fibronectin is composed of
two identical or nearly identical subunits linked through a disulfide bridge near the carboxyl
end (Figure 11). The combined molecular weight of the dimer is 450,000.

A similar molecule found in the human plasma, called cold insoluble globulin, is known
to be generally similar to fibronectin with small differences.[57] Asparagine-linked carbohydrate chains are present on the fibronectin molecules. The interaction of fibronectin molecules

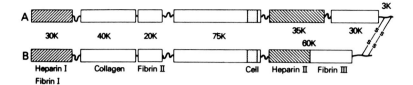

FIGURE 11.   Functional domains of fibronectin molecule. The C-terminals of the two subunits are at the right ad the N-terminals at the left. The binding specificities of different polypeptide fragments and their molecular weights are indicated. (From Yamada, K. M., Hasegawa, T., Hasegawa, E., Kennedy, D. W., Hirano, H., Hayashi, M., Akiyama, S. K., and Olden, K., *Prog. Clin. Biol. Res.*, 151, 1, 1984. With permission.)

among themselves and with other components of the extracellular material is interesting and has been investigated only recently. An important reason for the initial interest in fibronectin was the observation that its amount is considerably decreased after cell lines are transformed by oncogenic viruses and carcinogens. Tumor cell lines also have greatly decreased levels of fibronectin. Based on this observation, the cell surface fibronectin was called the LETS protein (Large External Transformation Sensitive protein). The decrease is more than tenfold in transformed human fibroblasts and glial cells. Whether there is any causal relation between the decreased levels of fibronectin and tumorigenicity is, however, not clearly understood. Besides, the biological role of fibronectin has now been shown to be much wider.

Soluble fibronectin molecules can assemble spontaneously into filamentous structures. Polyamines promote filament formation. A striking feature of the fibronectin molecules is that it has different distinct domains for binding various substances. At the N-terminal end of the polypeptide, there is a domain containing binding sites for fibrinogen/fibrin and actin. Binding of fibrinogen or fibrin is covalent and occurs through a glutamine residue. A weak heparin binding activity has also been demonstrated in this domain. The next region has binding sites for collagen. Sulfated polysaccharides such as heparin can enhance the rate of binding of fibronectin to collagen.[58] Most of the carbohydrate (an average of five asparagine-linked oligosaccharides per subunit) is located nearer the C-terminal of this domain. The role of these oligosaccharides is not clear, since fibronectins devoid of them are as active as the glycosylated molecules. It is possible, however, that the carbohydrates provide stability and protection against degradation by proteolytic enzymes. Alternatively they could serve to give the molecule its ''normal'' shape relevant to building the matrix. Next to the collagen-binding region is the cell surface attachment domain. A glycosaminoglycan binding domain is located close to the C-teminal end of the polypeptide. This domain also has strong affinity for heparin. Heparin and glycosaminoglycan binding is noncovalent. The various domains of the fibronectin molecule are illustrated diagrammatically in Figure 11.

The exact mechanism of cell surface-fibronectin attachment is far from clear. Soluble fibronectin has low affinity for the cell surface, unlike the insoluble form such as that found in the extracellular matrix. There is only one attachment site per polypeptide close to the C-terminal and flanked by a heparin binding domain. Cell attachment affinity has been localized to a tetrapeptide, Arg-Gly-Asp-Ser. This fragment can inhibit competitively in fibroblast-immobilized fibronectin binding assays.[59] Essentially the assay procedure utilized CH Sepharose® 4B beads (Pharmacia, Sweden) on which are fibronectin molecules bound covalently. Fibroblasts added to such beads bind specifically on the fibronectin molecules. This binding can be inhibited competitively by the addition of free fibronectin or the test peptides.

An interesting fact revealed in these studies is that the same sequence is present in fibrinogen (or fibrin) and some other less relevant proteins. This domain of the fibrinogen molecule has, however, very poor capacity to bind the surface of fibroblasts. On the other hand, a nona-peptide with a slightly different sequence, Arg-Gly-Asp-Thr-Gly-Ala-Thr-Gly-Arg, taken from Type I collagen shows cell binding affinity almost as much as the fibronectin fragment. It has been pointed out that the neighboring sequences on either side of the tetrapeptide fragment are different in fibrinogen and fibronectin, and this is likely to be the basis of the different binding affinities. Obviously we do not know enough about the mechanism of the fibronectin-cell surface binding process. It is, however, gratifying that pursuing the studies such as those described above can pave the way towards such a goal.

What is the nature of the cell surface domain on which fibronectin binds? It has been shown that negatively charged lipids such as glycolipids containing several sialic acid residues are the binding sites. The implication of this is clear: fibronectin binds directly on the plasma membrane without the mediation of link molecules, and in the relatively ''naked'' regions devoid of peripheral proteins. Another glycoprotein that binds specifically to hyaluronic acid has been described.[60] It has been named hyaluronectin. It is secreted into the extracellular spaces. This glycoprotein is distinct from fibronectin. Not much is known about it so far.

Literature on fibronectins is already voluminous. For details and references, the following papers may be consulted: Yamada and Olden;[57] Pearlstein et al.;[61] Vaheri and Alitalo;[62] Ruoslahti et al.;[63] Klebe and Mock;[64] and Yamada et al.[65]

b. Laminin

Laminin is a recently discovered glycoprotein that serves as an important structural component of the basal lamina. It is distinct from fibronectin and collagen by criteria such as amino acid composition, immunological cross-reactivity, and electrophoretic mobility. The molecular weight of a subunit of laminin is 220,000. It is rich in cysteine. Extensive cross-linking of the subunits through disulfide bridges leads to the organization of the structural scaffolding of the basal lamina. Laminin is richly glycosylated (amount of carbohydrate 12 to 15%) with sialic acid as a significant part. Laminin shows binding affinity for heparin, cell surface, and Type IV collagen. It is secreted by epithelial cells and by fibroblasts.

Information on laminin is scanty, unlike in the case of fibronectin. It is, however, clear that laminin is an adhesive glycoprotein for epithelial cells. For initial studies on laminin, reference may be made to Timple et al.[66] More recent information on the molecular structure and other aspects may be obtained from Terranova et al.[67] and Rao et al.[68,69]

c. Chondronectin

Chondronectin is yet another glycoprotein of the extracellular matrix, identified recently. It mediates attachment between chondrocytes and collagen Type II. It is present in serum and in cartilage. Very little information is available on the chemistry of this glycoprotein.[70]

E. Assembly of Matrix Components

The components of extracellular matrix described above are organized into complex structures constituting the acellular part of the connective tissues. The constituents are products elaborated in the cells and discharged into their environment. The diversity of the material can be the basis of different structures of the various connective tissues in the body.

The physicochemical attributes of the secreted molecules may be an important aspect of the mechanism of their ordered assembly. The most relevant attributes here are their solubility, charge, hydrophilic/hydrophobic character, and the ability to form covalent and non-covalent linkages with each other. Being a multistep process involving the assembly of more than one kind of component, the sequence of their elaboration in the cells and timed discharge could effectively control the form of the final product. One can also speculate that new attributes may arise in the assembled matrix which in turn could control the cellular activities such as adhesion and migration. In vitro assembly of matrix components has been used as a tool to reveal the physicochemical consequences of their molecular structure to the resulting product. From such studies, it has been concluded that much of the assembly process is driven by the physicochemical attributes of the components resulting in a near-normal organization in vitro. Collagen molecules assemble into fibrils spontaneously at neutral pH. Such fibrils have physical properties similar to those possessed by the fibrils formed in vivo.

Processing of the matrix components may start even before secretion. In some cases, the processing is already evident in the cells that secrete them. Rod-like structures, presumably collagen, are detectable in some of the Golgi vesicles of certain cells. Even the extracellular assembly of collagen and other material may be controlled by the cells. Infoldings of chick tendon cell surface from which organized collagen strands are discharged have been discribed.[71] The cell can thus control the length and orientation of the fibrous bundles. We shall later have an occasion to consider the importance of three-dimensional organization of the extracellular matrix in morphogenesis (Chapter 8, Volume II).

Although not much is known at present about the detailed control mechanisms of matrix assembly in the extracellular spaces, the existence of definite organization justifies the speculation that "prefabricated" smaller units consisting of macromolecular assemblies may be discharged from the cells. Further assembly of the "prefab" units may be modulated by their enzymic modifications, and higher structural organization may occur as dictated by their physicochemical properties. The massive structures such as tendons and the annulus fibers of intervertebral disc are far from haphazard deposits of the prefab units. Birefringence in connective tissue matrix clearly points to this fact. (Birefringence is the separation of a ray of light into two unequally refracted polarized rays and is caused by oriented deposition of fibrous molecules). Experimental studies on model assemblies of matrix molecules in vitro under various simulated conditions of in vivo milieu can throw considerable light on the mechanism of the process.

F. Regional Specialization of the Plasma Membrane

It seems from several lines of evidence that the cell adhesive mechanism is not a single general feature of all cells making them sticky. When one examines the various aspects of the problem, it becomes obvious that the precise structural relationships established during the differentiation of cells must depend on intricate adhesive mechanisms. Whereas fibronectin is associated with the adhesion of fibroblasts to collagen, laminin is involved in the adhesion of epithelial cells. An interesting difference in the behavior has been observed between Ewig's sarcoma and colon carcinoma cells. Adhesion of the sarcoma cells to plastic surface is induced by fibronectin, whereas that of carcinoma cells is by laminin.[22] This suggests that the various cell types have unique adhesive mechanisms.

Let us now consider the example of a hepatocyte *in situ*. The cell surface has three different types of connections with its environment. Over a certain area of the surface, it has another hepatocyte closely apposed to it with peculiar junctional specializations. The region facing a sinusoid is in contact with a reticular fiber with collagen, and the region facing the bile canaliculus is the free luminal surface of the cell (Figure 12). Ocklind et al.[72] posed the question: is the cell adhesion mechanism involved in cell-cell contacts different from cell-collagen contact? From the hepatocytes the plasma membrane fragments at the sinusoidal surface (cell-collagen contact) and at contiguous faces of neighboring hepatocytes (cell-cell contacts) were obtained separately. (See Wisher and Evans[73] and Cook et al.[74] for the more recent methods of obtaining the plasma membrane domains.) Using these enriched fractions as antigens, antibodies were raised in rabbits. Adhesion of hepatocytes mutually in reaggregation and to collagen in vitro was studied in the presence and absence of such antibodies. Antibodies against "cell-cell adhesion molecules" prevented mutual adhesion of hepatocytes, but not their adhesion to collagen. Conversely, antibodies against "cell-collagen adhesion molecules" inhibited adhesion of hepatocytes to collagen, but not to other hepatocytes.

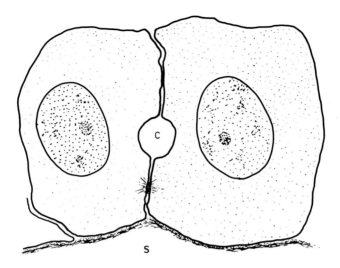

FIGURE 12.   Different domains of the plasma membrane on the surface
of hepatocytes. The diagram shows two cells adhered together. The plasma
membrane domains exposed to the lumen of a canaliculus (C), to a sinus
(S), and to the plasma membrane of another cell may be distinguished.

These results show clearly that the adhesion molecules at the apposed plasma membranes
of two cells are different from those at cell-collagen contacts. Recent work has confirmed
microheterogeneity of the cell surface in hepatocytes[73] and the capillary endothelial cells.[74]
The implications of these findings to development and differentiation of embryonic tissues
are obvious. When cells assume definite mutual relations with their cellular and noncellular
environment, the mechanism leading to such positioning may well be the appearance of
specific regionally differentiated adhesive surfaces.

IV. MECHANISMS OF CELL ADHESION

A. Formation of Cell Junctions

Ultrastructural studies have established that cells adhere to each other through the mediation
of different structural specializations of the plasma membrane. Many tissue cells are separated
widely due to the presence of intercellular substances. However, even the cells that are
closely apposed to each other are separated by a gap measuring 100 to 200 Å of electron
light region over a large part of the contact area.

There are, however, specialized regions of cell contact such as the desmosomes, tight
junctions, and gap junctions (Figure 13). The desmosomes (also known as adherens junctions
or macula adherens) are round to oval regions of attachment of adjacent plasma membranes
where they are separated by a gap of 250 to 350 Å. An electron dense material fills the
gap, with a specially dense central plane. On the cytoplasmic face of the membrane is an
amorphous plaque which is the site of attachment of tonofilaments.[75] The attachment of the
plaque to the membrane has been shown to be mediated by α-actinin in case of cultured
fibroblasts.[76] Proteins isolated from the desmosomes contain a high proportion of nonpolar
amino acid residues, suggesting that these proteins are anchored in the lipid bilayer. Structures
called hemidesmosomes are those that are asymmetrical, i.e., have a plaque only on one
side of the junction.

In another type of specialized zone of membrane contact, viz., the tight junctions (also
known as occludens junctions or zonulae and maculae occludens), the plasma membranes
of adjacent cells are so close that the intercellular space is excluded. Usually the tight

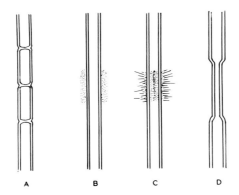

FIGURE 13.   Diagram to show the different types of cell junctions. A, zonula occludens; B, zonula adherens; C, desmosome; D, gap junction.

junctions are linear domains that in some cases may form a continuous belt around the cell. Gap junctions are regions of membrane contact where the closely apposed plasma membranes of two cells are coupled by cylindrical protein aggregates. These regions are permeable to small ions carrying electric current. It is not intended to give a detailed description of the junctional specializations here. For details, see Staehelin.[77] An important point to be emphasized here is that the different theories on the mechanism of cell adhesion should be in accord with the ultrastructural observations.

1. The Biophysical Aspects of Cell Adhesion

Curtis[78] provided a detailed discussion on the biophysical forces that could hold cells together separated by a 100 to 200 Å space. More recent discussions of the subject are presented by Doroszewski,[79] Gingell and Vince,[80] and Rutter.[81] According to Curtis, the weak reversible adhesive forces between colloidal particles operate in cell adhesion. More intimate junctions (tight junctions) are formed only if the repulsive energy barrier is crossed by the approaching cells, and molecular bridges are formed. One should guard against gaining the impression that cells adhere to others by a single mechanism all over the surface. If a uniform surface is assumed (as is done in most biophysical treatments of the problem), the overall shape of cells in an aggregate, such as a tissue, would always be somewhat isodiametrical (or, more accurately, orthotetrakeidecahedron, the shape assumed by a soap bubble surrounded by other bubbles from all sides), which is far from true. The inevitable conclusion is that such approximations are only of theoretical interest in providing a conceptual framework for some of the observed facts. If taken literally, however, they are not only useless, but even misleading.

An important consideration in the discussions on the biophysics of cell adhesion is the surface free energy. Stable adhesion is favored by minimizing the surface free energy. The surface of a substance or boundaries between phases (solid/liquid, liquid/liquid, etc.) requires one extra term in the description of its energy, which is due to its location at the boundary where a sharp change in the concentration or properties of the substance occurs. The energy associated with this location is the free surface energy. In case of liquids, this is analogous to surface tension. The surface free energy is a consequence of the molecular properties of the materials. In case of cells, the molecules constituting the plasma membrane confer the surface free energy. The intermolecular forces can be categorized as follows:[78] (1) Van der Waals forces consisting of attraction between permanent dipoles, dipoles induced by other permanent dipoles, and "statistical dipoles" resulting from the random motion of electrons in nonpolar materials; (2) hydrogen bonds. Evidence from the studies on the electrophoretic behavior of cells indicates that the cells bear a net negative charge at physiological pH, and

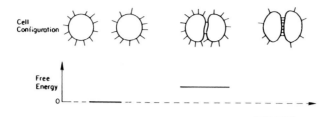

FIGURE 14. Changing energy level during the process of establishing cell contact. When two cells first come into contact, the free energy is increased (*center*) and only after the receptors or adhesion molecules have accumulated in contact area (*right*) is the free energy below that of the separated cells. (Reproduced from *Biophys. J.*, 1984, 45, 1051, by copyright permission of the Rockefeller University Press.)

this could create an important repulsive barrier between cells. The existence of a repulsive barrier between cells is further emphasized by the studies showing that even when the cells are capable of adhesion by virtue of possessing adhesion molecules, they must frequently be forced into a close approximation by centrifugation before strong bonding occurs.[82] The repulsive force can be counter-balanced by the establishment of specific bonding between adhesion molecules of the cells. Rearrangement of the adhesion molecules in the membrane may be an essential prerequisite of the adhesive process. The free energies of the cells in the process of adhesion are depicted in Figure 14. The thermodynamic analysis of the process is complex. Interested readers are referred to the article by Bell et al.[83]

2. The Metabolic Aspects of Cell Adhesion

We shall now consider the molecular mechanisms by which cells could be held together. In a previous section we have already seen how Ca^{2+} and Mg^{2+} ions are important in holding cells together. The precise role of these divalent cations is, however, not very clear. An old idea was that they can form bridges between anionic binding sites (Figure 4). This could occur directly between two plasma membranes or via a connecting molecule. Calcium ions may bind the carboxyl groups of amino acid residues, sialic acids, or to the phosphate groups. In order to form a direct Ca^{2+} bridge between two cells, they will have to be held very close together. Since this cannot be reconciled with the observed 100 to 200 Å space between plasma membranes, a large cementing molecule or molecular complex will have to be postulated. As an alternative explanation, it has been suggested that Ca^{2+} tends to lower the surface negative charge so that the mutual repulsive force between cells decreases. If this were so, different cations could have a similar action on cell adhesion. Obviously it is not so. Armstrong[84] has shown that magnesium is more effective than calcium in bringing about aggregation. It seems that not enough is known regarding the mechanism to explain the role of divalent cations in cell adhesion.

It has been observed that disaggregation of cells in Ca-Mg-free media results in the liberation of proteins (and/or glycoproteins), which can promote cell reaggregation. It is also known that cells cultured in vitro secrete considerable amounts of proteins capable of promoting cell reaggregation. This may be a normal response of the cells to the absence of such proteins around them. It is conceivable that a cell tends to secrete these "cementing" molecules continuously when the inhibitory feedback of normal cellular environment established through the ligands is absent. The cementing proteins can form a bridge between two plasma membranes causing them to be held together.

Recently Jones[85,86] has proposed an elaborate mechanism of the formation of adhesive contacts through the development of desmosomes. There are distinct stages in the development of the symmetrical adhesive complexes, viz., the desmosomes. An initial event in

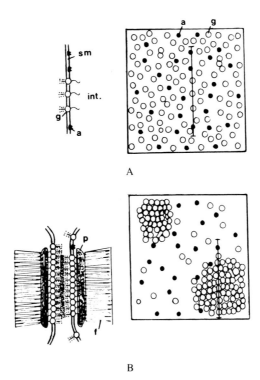

FIGURE 15.   Formation of cell adhesion plaque complex in response to clustering of transmembrane glycoproteins in a region of cell-cell contact. The squares represent freeze-fracture replicas. On the left of each square is a part of the surface membrane corresponding to that part marked by a vertical line in the square. (A) glycoproteins (g) and enzyme systems (a) are uniformly distributed. (B) Mobile glycoproteins have clustered in regions of cell-cell contact and have displaced and forced the enzyme systems out of the region of contact. In response to these changes a plaque is formed; note the apposed clusters of glycoproteins with oligosaccharides in the interspace and the side-by-side microfilaments in the plaque (p). f, Bundle of microfilaments aligned in the cytoplasm; int, interior of the cell; sm, cell surface membrane. (From Jones, B. M., Regulation of the contact behaviour of cells, *Biol. Rev.*, 55, 207, 1980. With permission of the publisher, the Company of Biologists, Cambridge University Press.)

the process is the appearance of a dense material in a submembrane location, assuming the form of a plaque. A dense mass of actin filaments parallel to the membrane is then formed. Organized filamentous material containing glycoproteins then appears in the intercellular space. It is also known from freeze-fracture studies that transmembrane glycoproteins cluster at such cell contact sites. Cytoplasmic tonofilaments then orientate towards the plaque. The adhesive complex thus formed is now a desmosome. The scheme proposed by Jones[85] postulates that localized changes in the membrane involving the distribution of adenylate cyclase and the calcium pump ATPase bring about the structural changes leading to the formation of the desmosomes. It is also proposed that clustering of the glycoproteins causes displacement of other nonjunctinal proteins outside the contact area (Figure 15). The non-junctional proteins displaced from the plaque-forming area include adenylate cyclase system, and consequently there is a reduction in the amount of cyclic AMP (cAMP) produced. However, the cAMP continues to be degraded on the cytoplasmic face by soluble phosphodiesterase. Thus there is a localized fall in the cAMP.

   Another link in this rather complex set of events is a Ca^{2+}-dependent modulator protein, which at low calcium level does not activate the phosphodiesterase. When calcium is abundant, the modulator protein interacts with phosphodiesterase that degrades cAMP. The Ca^{2+}-dependent protein can also activate myosin-ATPase, which then promotes the transformation of globular actin (G-actin) to filamentous actin (F-actin). The actin filaments form the plaque in the domain of the adhesive contact where glycoproteins have clustered. Jones[85] has proposed the following sequence of interactions leading to the rapid assembly of actin filaments:

$$Ca^{2+} + \text{modulator protein (MP)} \rightleftharpoons Ca^{2+} MP$$

$$Ca^{2+}MP + \text{myosin (My) ATPase} \rightleftharpoons Ca^{2+} MP.My\ ATPase$$

$$\text{G-actin ATP} \underset{\xleftarrow{\hspace{3cm}}}{\overset{Ca^{2+}MP.My\ ATPase}{\xrightarrow{\hspace{3cm}}}} \text{F-actin ADP}$$

(Mg^{2+} is responsible for release of phosphate from actin filaments.)

By analogy with muscle, it is assumed that when cAMP level decreases, there is an increase in the level of cytoplasmic Ca^{2+} (released from other storage structures). This localized increase in the level of Ca^{2+} starts the interactions outlined above. The structural reorganizations involved in the above process are represented diagrammatically in Figure 15.

The physicochemical model put forward by Jones[85,86] rests on several assumptions. However, if future work vindicates the assumptions, it would constitute an attractive model for the establishment of desmosomes at cell adhesive contacts. There seems to be no relation between this scheme of cell adhesion and the Ca^{2+}-dependent mechanism described by Takeichi et al.,[33] Grunwald et al.,[35] and Magnani et al.,[37] which has been mentioned earlier. The mechanism described by these workers seems to consist of some molecules exposed on the cell surface and related to the initial aggregate formation in a cell suspension. There is no evidence that the putative calcium dependent mechanism described by these workers directly promotes desmosome formation. It is of course possible that there are several mechanisms of cell adhesion, and their relative importance may be variable according to the cell types. Obviously, not all cell adhesions occur exclusively through desmosomes. The time taken for the establishment of various types of junctions is variable. Desmosomes take 90 min to form in a cell line obtained from human cervical cancer.[87] On the other hand, tight junctions take a shorter time ($\approx$30 min in *Necturus* gall bladder epithelium), and gap junctions take still shorter time.[88,89] It must be noted that the Ca^{2+}-dependent mechanism described by Takeichi et al.[33] and other workers[34-39] has been demonstrated in the reaggregation of cells in stirred suspensions. In all probability, this does not involve the formation of desmosomes, which seems to be a much later event.

It may be assumed, albeit *a priori,* that all the molecular components of the plasma membrane could have some direct or indirect role to play in cell adhesion. Curtis et al.[90,91] altered the composition of membrane lipids by experimental incorporation of fatty acids. Incorporation of long chain length saturated fatty acids was found to increase the adhesiveness of cells, whereas incorporation of unsaturated fatty acids caused a fall in adhesiveness. A certain degree of fluidity of the plasma membrane is essential for all the functions performed by this organelle, and it may be hypothesized that modulation of the property may be effected by physiological alterations in the lipid composition of the membrane. There is, however, no experimental evidence to elaborate such a hypothesis further.

B. Cell Adhesion Molecules

Recently a wide variety of proteins and glycoproteins of the plasma membrane have been assigned various roles in cell recognition and adhesion. No generalizations can be made from the extant knowledge. However it will be useful to describe some of the well-investigated examples so as to give an idea regarding the complexity of cell adhesive mechanisms. It must be emphasized that extant information on the cell adhesion molecules is of only recent origin. It is likely that many different adhesion molecules will be described in the future. These may be tissue specific or shared by several anatomically related tissue types.

1. Neural Cell Adhesion Molecule

Edelman and associates[92,93] have studied the "neural cell adhesion molecules" (N-CAM) from chick neural retina and other neural tissues. These are located on the surface of the plasma membrane and are involved in cell adhesion. Monovalent Fab against the cell adhesion molecule can inhibit reaggregation of neural retina cells effectively and also inhibit histogenesis of neural retina in organ culture.[95] Neural retina cells from 8-day embryos aggregate about four times faster than older ($\approx$14 days) cells, and this accords with a comparable abundance of the adhesion factor on the surface of the younger cells. The estimated molecular weight of the cell adhesion molecule isolated by Rutishauser et al.[92] is 140,000. Grumet et al.[96] have demonstrated that chick embryo skeletal muscle cells also have cell adhesion molecules characteristic of the neural tissue. They have also demonstrated that in vitro adhesion of the skeletal muscle cells to neural retina cells can be inhibited by antibodies against the N-CAM. The interaction of nerves with muscles, particularly at the neuromuscular junctions, is very specific, and these are probably established through the mediation of the cell adhesion molecules.[97]

Edelman and colleagues[93] isolated the N-CAM from the neural retina, brain, and muscle of chick and other species. An important step in the purification of the N-CAMs is raising monoclonal antibodies specific for a characteristic part of the molecule (e.g., sialic acid of the N-CAM) and using the antibodies in affinity columns. This method has enabled purification of the N-CAMs for structural studies. It has been shown that the N-CAMs occur as integral membrane components. Structurally they consist of a single polypeptide chain with variable amounts of carbohydrate (26 to 35%). Owing to the variation in the carbohydrate content, the N-CAM shows a microheterogeneity in electrophoresis. Almost four fifths of the carbohydrate are represented by sialic acid, which is strikingly higher compared with any other surface glycoproteins. In aqueous solutions, the N-CAMs show self-aggregation. This must be due to the hydrophobic domains of the molecules sequestering away from the aqueous phase. This property offers difficulties in quantitative assessment of their binding on the cell surface.

Preparation of artificial lipid vesicles containing the N-CAM has partly overcome these difficulties. Such lipid vesicles bind specifically on the surface of cells of neural retina, brain, and myoblasts/myotubes. These vesicles also aggregate themselves. This aggregation as well as N-CAM-dependent aggregation of the cells is Ca^{2+}-independent. Fab fragments of antibody to N-CAM inhibit specifically the binding of vesicles on the cell surface. The NH$_2$-terminal domain of the polypeptide of a molecular weight $\approx$65,000 seems to bear the binding site, as inferred from the observation that this fragment inhibits binding of the N-CAM to the cells. Neuraminidase-treated N-CAMs do not lose binding affinity; in fact, binding is even more effective. From this, it has been concluded that the sialic acid is not located at the binding sites. The heterogeneity in the N-CAMs sialic acid content has been shown to influence its binding properties as desialylation increases the rate of its binding to cells by an order of magnitude. This could be a mechanism by which the degree of adhesiveness of cells can be controlled. Cell-cell adhesion through the N-CAM is illustrated in Figure 16. It must be noted that the exact mechanism of adhesion will be clarified only after more structural information about the N-CAM is available. However, the investigations by Edelman and colleagues described above have indicated clearly that this mechanism of cell adhesion is based on identical or similar complementary molecules located at the cell surface, and tissue specificity is determined by small variations in the nature of the adhesion molecules. Besides, ontogenic changes in the adhesion molecules (especially through glycosylation) can modulate their function. A definite change has been shown to occur in the N-CAM as the embryo develops into the adult. The difference between the embryonic and adult N-CAMs is chiefly with regard to their sialic acid content.[98] The role of N-CAM in synapse formation and other developmental changes has been elucidated using certain re-

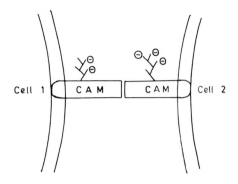

FIGURE 16.   Diagram to explain cell adhesion mediated by the neural cell adhesion molecule as proposed by Edelman.[93] Branching oligosaccharides away from the domain of adhesion are shown with negative signs representing sialic acids. See the text.

cessive mutants in the mice in which these processes are defective. For further details and references, see Edelman[93,99] and Rutishauser.[94]

2. Cognin, Adherons, and Others

A cell surface glycoprotein functionally similar to that described by Rutishauser et al.[92] was obtained from the neural retina of 10-day chick embryo. The glycoprotein has been named retina cognin by Ben-Shaul et al.[100] It has been purified extensively and its molecular weight is 50,000. Cognin is responsible for specific recognition of retina cells. The presence of cognin as a plasma membrane component has been demonstrated by a technique that involves raising antibodies against it in rabbits. The antibodies bind the membrane cognin in intact cells. These cells are now exposed to polystyrene microbeads (average diameter 0.23 μm), which are coated with goat antirabbit antibodies. The goat antibody-coated beads would thus bind only the cell surface sites where the anticognin antibodies have bound. The microbeads are large enough to be shown by scanning electron microscopy (Figure 17).

Nonneural cells do not possess cognin in their membranes, suggesting tissue specificity. A small proportion of some other neural cells have cognin in their membrane. The membrane location of cognin has also been shown by other lines of evidence. Trypsinized cells do not show labeling of microbeads as described above unless the cells have recovered from the trypsin trauma. Anticognin can cause complement-mediated immunolysis of neural retina cells. These facts indicate that cognin is a true plasma membrane protein. It seems to have organ specificity extending beyond the animal type as known from the observation that chick cognin promotes aggregation of mouse retina cells. Do the cognin molecules bind directly with other cognin molecules of an apposed cell? Alternatively, is there a linking mechanism required for binding? The requirement of such a linking molecule has in fact been described by Rutz and Lilien[101] (see Figure 18). Another very important question that needs to be answered is: are the "cognin" and N-CAM related in any way? No definite information is yet available to answer these questions. However, extant biochemical technology is adequate to obtain such information, which will enhance understanding of the mechanisms of cell adhesion.

Adheron is the name given to a macromolecular complex consisting of a limited number of proteins — collagen, fibronectin, and glycosaminoglycans.[102-104] This complex is released by cells into the culture media, especially those lacking serum. Adherons released by skeletal muscle cells sediment at 16S. Somewhat similar molecular complexes (12S) have been obtained from the neural retina, and other tissues. These seem to be tissue specific as shown by experiments. Adhesion of neural retina cells to petri dishes coated with putative adherons

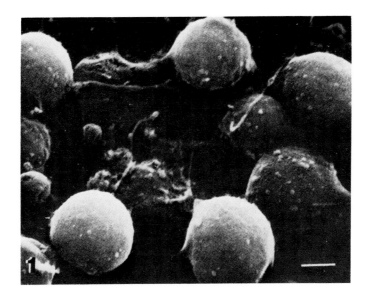

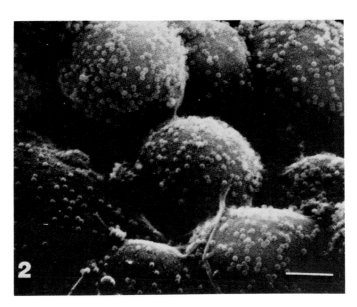

FIGURE 17.   Visualization of the neural retina cognin. In the upper figure are cells treated with anticognin antibody to which polystyrene microbeads are bound. See the text. In the lower figure are cells trypsinized before exposing to the anticognin antibody. Note the absence of microbeads on the cells indicating the absence of the antigen (cognin) on the cell surface. (From Ben-Shaul, Y., Hausman, R. E., and Moscona, A. A., *Dev. Biol.*, 72, 89, 1979. With permission.)

from neural retina, myoblasts, and pigmented retina was compared. Maximum adhesion was observed on petri dishes with the neural retina adheron. It is, however, not certain if the difference is inherent in the adhesion mechanism or is due to variation in the composition of the molecular aggregates. Obviously more information on the adheron is needed.

It has been shown that the neural retina adheron shares antigenic determinants on the cell surface, suggesting that at least some part of the adheron is a membrane protein. It is likely, however, that the adheron is a complex formed in a manner dictated by the physicochemical

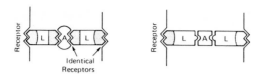

FIGURE 18.   Cell adhesion mediated by cell surface receptors and link molecules. Two alternative schemes are shown. On the left is a mechanism in which a divalent adhesion molecule (L) has a binding site for the cell surface and a similar site which may bind another linker molecule (A). On the right is shown a mechanism in which the adhesion molecule (L) has two distinct binding sites, one for the cell surface and another for a linking molecule. (From Rutz, R. and Lilien, J., Functional characterization of an adhesive component from embryonic chick neural retina, *J. Cell Sci.*, 36, 323, 1979. With permission of the publisher, the Company of Biologists, Cambridge University Press.)

properties of several functionally related cell adhesion molecules released into the culture medium. It will therefore be interesting to know more about the manner in which adherons are assembled in vitro. Such information may eventually lead to a better understanding of the manner in which several macromolecules act in unison to bring about the complex tissue architecture.

Adheron-mediated adhesion of neural retina cells is calcium independent. In this respect, it resembles the N-CAM described earlier. A calcium-dependent adhesion molecule has been obtained from a murine mammary tumor cell line and mouse teratocarcinoma cells.[105,106] This is a membrane glycoprotein named "cadherin". Its molecular weight is $\approx 124,000$.[40] An 84,000 mol wt glycoprotein isolated from serum-free conditioned medium of mouse mammary tumor epithelial cells seems to be a fragment of the larger 124,000 mol wt membrane molecule. This seems to be widely distributed in several epithelia as a cell-cell adhesion molecule. Monoclonal antibody (ECCD-1) against this Ca^{2+}-dependent adhesion molecule can cause decompaction of mouse blastocyst cells at 8 to 16 cell stage[106] (see Chapter 8 for a description of the compaction of mouse morula cells). The antibody (ECCD-1) can also induce disruption of cell adhesion in cultures of an epithelial cell type derived from mouse teratocarcinoma. It also prevents the aggregation of these cells. It binds certain epithelial cells such as hepatocytes, epidermal cells, and alveolar cells of lung, but not some other epithelia. It does not bind at all to fibroblasts and cells derived from embryonic brain. Only cell-cell adhesion, and not cell-substratum adhesion is affected by the antibody.[40] Other calcium-dependent mechanisms have also been reported; they seem to promote cell-substratum adhesin. Whether and how they are related to "cadherin" is, however, not clear.

3. Holding the Cells Together in Tissues

As more and more "new" cell adhesion molecules and their complexes are discovered,[107,108] it is necessary to integrate the information so as to be relevant to understanding cellular organization in tissues. The exact manner in which cell adhesion molecules bring about cell contacts is only beginning to be understood. In case of a mechanism wherein only one molecular species is involved, adhesion may occur directly or through a linking molecule. For the N-CAM, it has been shown that there is no linking molecule. Integral membrane proteins/glycoproteins released into aqueous media tend to form oligomers by binding at the hydrophobic domains. Such oligomers can then promote aggregation of cells with intact membrane-located adhesion molecules (Figure 19A). In case a link molecule is involved, more than one kind of molecular binding domains have to be assumed (Figure 19B).

Cell adhesion in the formation of tissues is of two distinct types: cell-cell adhesion and cell-extracellular matrix adhesion. Since several components of the matrix are involved, there must be multiple mechanisms of cell-matrix adhesion. The normal shape of the cell could also be a consequence of the nature of the adhesion mechanism. Aplin et al.[109] have

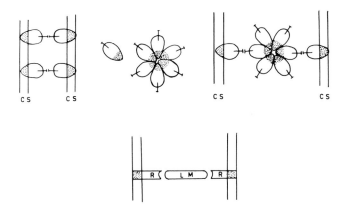

FIGURE 19.   Diagram illustrating two mechanisms of cell adhesion. The upper part illustrates the case of cell adhesion molecules (ovate outlines) which directly link two cells through their binding regions shown by T-shaped lines. These molecules are integral membrane components, and when released from the membrane into an aqueous medium, tend to aggregate through their hydrophobic regions (dotted parts) turned away from the bulk medium. The oligomers formed in this manner can also bind the adhesion molecules located in the intact plasma membrane (CS) and thus bring about cell adhesion. The lower part of the diagram illustrates a case in which adhesion molecules in the plasma membrane have receptors (R) which need a link molecule (LM) to bring about cell adhesion. In this case also the cell adhesion molecules would tend to aggregate into oligomers when dislodged from the plasma membrane. However, they would not be expected to cause cell adhesion unless the link molecules are also present.

presented experimental evidence to show that both fibronectin and laminin exert an effect on the amnion epithelial cell morphology. The shapes in the presence of these two substances differ from the shape when cells adhere to native matrix. Thus the two matrix components seem to act synergistically. In the native matrix, there are presumably several components, and their influence on the cells would be correspondingly complex.

V. CONCLUDING REMARKS

In this chapter, we have reviewed some of the recent work on the problem of cell adhesion. In the subsequent chapters we shall have occasion to discuss specificity of cell adhesion and its role in development. It is now necessary to consider only the major generalizations which seem to have wide acceptance among workers in the field. Attempts at quantitative estimations of the strength of cell adhesion have shown one thing: the methods developed so far are at best empirical. Different methods estimate different properties that are only indirectly related to the mechanism of cell adhesion. None of the methods seems to estimate the adhesive strength of cell-cell contacts in vivo. As pointed out earlier, it would be more profitable to determine the adhesive strengths of distinct molecular mechanisms rather than the ill-defined "cell adhesion" that is only a microscopically demonstrated end result of a complex and perhaps multicomponent process. It is also clear that the in vivo mechanisms may be specific as well as nonspecific. Cell adhesion in vivo is probably brought about by multiple mechanisms, which may be distinct from the nonspecific or less specific ones that cause clustering of cells in agitated suspensions.

Biophysical studies that have presented various models of possible molecular mechanisms of cell adhesion hold considerable promise of future progress. An urgent need in this context is to integrate the theoretical models with actual examples of cell adhesion. Recent studies reported by Bell et al.[83] have made such an approach. The most attractive feature of the biophysical models is that they provide a powerful method of analyzing and integrating various notions about the membrane-resident adhesion molecules (e.g., heterogeneity, density, and lateral mobility in the plane of the plasma membrane),[110-112] notions about the linking molecules (dimensions, compressibility, and other elastic properties and binding constants), notions about surface free energy, charge density, attractive and repulsive forces, and notions about geometrical parameters (contact area, contact distances, etc.).

REFERENCES

1. **Weiss, L.,** Studies on cellular adhesion in tissue culture, *Exp. Cell Res.,* 71, 281, 1972.
2. **Kolodny, G. M.,** Effect of various inhibitors on readhesion of trypsinized cells in culture, *Exp. Cell Res.,* 70, 196, 1972.
3. **Grinnel, F. and Srere, P. A.,** Inhibition of cellular adhesiveness by sulfhydryl blocking agents, *J. Cell. Physiol.,* 78, 153, 1971.
4. **Ambrose, E. J.,** The movememts of fibrocytes, *Exp. Cell Res.,* Suppl. 8, 54, 1961.
5. **Grinnel, F.,** Cellular adhesiveness and extracellular substrata, *Int. Rev. Cytol.,* 53, 65, 1978.
6. **Curtis, A. S. G., Forrester, J. V., McInnes, C., and Lawrie, F.,** Adhesion of cells to polystyrene surfaces, *J. Cell Biol.,* 97, 1500, 1983.
7. **Kemp, R. B., Jones, B. M., Cunningham, I., and James, M. E. M.,** Quantitative investigation on the effect of puromycin on the aggregation of trypsin and versene dissociated chick fibroblast cells, *J. Cell Sci.,* 2, 323, 1967.
8. **Appleton, J. C. and Kemp, R. B.,** Effect of cytochalasins on the initial aggregation *in vitro* of embryonic chick cells, *J. Cell Sci.,* 14, 187, 1974.
9. **Curtis, A. S. G.,** The measurement of cell adhesion by an absolute method, *J. Embryol. Exp. Morphol.,* 22, 305, 1969.
10. **George, J. V. and Rao, K. V.,** Rate of aggregation of disaggregated cells, *Indian J. Exp. Biol.,* 15, 174, 1977.
11. **Curtis, A. S. G. and Greaves, M. F.,** The inhibition of cell aggregation by a pure serum protein, *J. Embryol. Exp. Morphol.,* 13, 309, 1965.
12. **Rao, K. V., George, J. V., and Truman, D. E. S.,** On the physicochemical basis of cell adhesion, *Curr. Sci.,* 42, 340, 1973.
13. **George, J. V. and Rao, K. V.,** The role of sulfhydryl groups in cellular adhesiveness, *J. Cell. Physiol.,* 85, 547, 1975.
14. **Bellairs, R., Curtis, A. S. G., and Sanders, E. J.,** Cell adhesiveness and embryonic differentiation, *J. Embryol. Exp. Morphol.,* 46, 207, 1978.
15. **Thomas, W. A. and Steinberg, M. S.,** A twelve-channel automatic device for continuous recording of cell aggregation by measurement of small-angle light-scattering, *J. Cell Sci.,* 41, 1, 1980.
16. **Curtis, A. S. G.,** On the occurrence of specific adhesion between cells, *J. Embryol. Exp. Morphol.,* 23, 253, 1970.
17. **Roth, S. A. and Weston, J. A.,** The measurement of intercellular adhesion, *Proc. Natl. Acad. Sci. U.S.A.,* 50, 974, 1967.
18. **Roth, S., McGuire, E. J., and Roseman, S.,** An assay for intercellular cell adhesive specificity, *J. Cell Biol.,* 51, 525, 1971.
19. **Koziol, J. A., Springer, W. R., and Barondes, S. H.,** Quantitation of selective cell-cell adhesion and its application to assays of species-specific adhesion in cellular slime molds, *Exp. Cell Res.,* 128, 375, 1980.
20. **Sieber, F. and Roseman, S.,** Quantitative analysis of intercellular adhesive specificity in freshly explanted and cultured cells, *J. Cell Biol.,* 90, 55, 1981.
21. **Curtis, A. S. G. and Forrester, J. V.,** Interactions between cells and with intercellular matrices, *Prog. Clin. Biol. Res.,* 151, 339, 1984.
22. **Vlodavsky, I., Lui, G. M., and Gospodarowicz, D.,** Morphological appearance, growth behavior and migratory activity of human tumour cells maintained on extracellular matrix versus plastic, *Cell,* 19, 607, 1980.
23. **Gospodarowicz, D. and Ill, C. R.,** Factors involved in supporting the growth and steroidogenic functions of bovine adrenal cortical cells maintained on extracellular matrix and exposed to a serum-free medium, *J. Cell. Physiol.,* 113, 373, 1982.
24. **Steinberg, M. S.,** The role of temperature in the control of aggregation of dissociated embryonic cells, *Exp. Cell Res.,* 28, 1, 1962.
25. **Moscona, A. A.,** Effect of temperature on adhesion to glass and histogenetic cohesion of dissociated cells, *Nature (London),* 190, 408, 1961.

26. **Edwards, J. G. and Campbell, J. A.,** The aggregation of trypsinized BHK21 cells, *J. Cell Sci.,* 8, 53, 1971.
27. **George, J. V. and Rao, K. V.,** Cell aggregation and its temperature sensitivity, *Indian J. Exp. Biol.,* 13, 470, 1975.
28. **Nath, K. and Srere, P. A.,** Effects of temperature, metabolic and cytoskeletal inhibitors on the rate of BHK cell adhesion to polystyrene, *J. Cell Physiol.,* 92, 33, 1977.
29. **Steinberg, M. S.,** On the chemical bonds between cells: a mechanism for type-specific association, *Am. Nat.,* 92, 65, 1958.
30. **Pethica, B. A.,** The physical chemistry of cell adhesion, *Exp. Cell Res.,* Suppl. 8, 123, 1961.
31. **Curtis, A. S. G.,** Cell adhesion, *Sci. Prog.,* 54, 61, 1966.
32. **Oppenheimer-Marks, N. and Grinnel, F.,** Calcium ions protect cell-substratum adhesion receptors against proteolysis, *Exp. Cell Res.,* 152, 467, 1984.
33. **Takeichi, M., Okazaki, H. S., Tokunaga, K., and Okada, T. S.,** Experimental manipulation of cell surface to affect cellular recognition mechanisms, *Dev. Biol.,* 70, 195, 1979.
34. **Urushihara, H., Okazaki, H. S., and Takeichi, M.,** Immunological detection of cell surface components related with aggregation of chinese hamster and chick embryonic cells, *Dev. Biol.,* 70, 206, 1979.
35. **Grunwald, G. B., Geller, R. L., and Lilien, J.,** Enzymatic dissection of embryonic cell adhesive mechanisms, *J. Cell Biol.,* 85, 766, 1980.
36. **Grunwald, G. B., Bromberg, R. E. M., Crowley, N. J., and Lilien, J.,** Enzymic dissection of embryonic cell adhesive mechanisms. II. Developmental regulation of an endogenous adhesive system in the chick neural retina, *Dev. Biol.,* 86, 327, 1981.
37. **Magnani, J. L., Thomas, W. A., and Steinberg, M. S.,** Two distinct adhesion mechanisms in embryonic neural retina cells. I. A kinetic analysis, *Dev. Biol.,* 81, 96, 1981.
38. **Thomas, W. A. and Steinberg, M. S.,** Two distinct adhesion mechanisms in embryonic neural retina cells. II. An immunological analysis, *Dev. Biol.,* 81, 106, 1981.
39. **Thomas, W. A., Thomson, J., Magnani, J. L., and Steinberg, M. S.,** Two distinct adhesion mechanisms in embryonic neural retina cells. III. Functional specificity, *Dev. Biol.,* 81, 379, 1981.
40. **Yoshida-Noro, C., Suzuki, N., and Takeichi, M.,** Molecular nature of calcium-dependent cell-cell adhesion system in mouse teratocarcinoma and embryonic cells studied with a monoclonal antibody, *Dev. Biol.,* 101, 19, 1984.
41. **Grinnel, F., Milam, M., and Srere, P. A.,** Studies on cell adhesion. II. Adhesion of cells to surface of diverse chemical composition and inhibition of adhesion by sulfhydryl binding reagents, *Arch. Biochem. Biophys.,* 153, 193, 1973.
42. **Grassetti, D. R.,** Preventing the spread of cancer, *Chemtech,* p. 666, 1973.
43. **Mehrishi, J. N. and Grassetti, D. R.,** Sulfhydryl groups on the surface of intact Ehrlich ascites tumour cells, human blood platelets, and lymphocytes, *Nature (London),* 224, 563, 1969.
44. **Rao, K. V.,** Irreversible inhibition of cell aggregation by some sulfhydryl reagents, *Curr. Sci.,* 42, 826, 1973.
45. **Ramachandran, G. N. and Ramakrishnan, C.,** Molecular structure, in *Biochemistry of Collagen,* Ramachandran, G. N. and Reddi, A. H., Eds., Plenum Press, New York, 1976, chap. 2.
46. **Eyre, D. R., Paz, M. A., and Gallop, P. M.,** Cross-linking in collagen and elastin, *Annu. Rev. Biochem.,* 53, 717, 1984.
47. **Kurkinen, M., Taylor, A., Garrels, J. I., and Hogan, B. L. M.,** Cell surface associated proteins which bind native type IV collagen or gelatin, *J. Biol. Chem.,* 259, 5915, 1984.
48. **Ramachandran, G. N. and Reddi, A. H., Eds.,** *Biochemistry of Collagen,* Plenum Press, New York, 1976.
49. **Bornstein, P. and Sage, H.,** Structurally distinct collagen types, *Annu. Rev. Biochem.,* 49, 957, 1980.
50. **Linsenmayer, T. F.,** Collagen, in *Cell Biology of Extracellular Matrix,* Hay, E. D., Ed., Plenum Press, New York, chap. 1.
51. **Piez, K. A. and Reddi, A. H., Eds.,** *Extracellular Matrix Biochemistry,* Elsevier, New York, 1984.
52. **Partridge, S. M., Elsden, D. F., and Thomsa, J.,** Constitution of the cross-linkages in elastin, *Nature (London),* 197, 1297, 1963.
53. **Franzblau, C. and Faris, B.,** Elastin, in *Cell Biology of Extracellular Matrix,* Hay, E. D., Ed., Plenum Press, New York, 1981, chap. 3.
54. **Senior, R. M., Griffin, G. L., Mecham, R. P., Wren, D. S., Prasad, K. U., and Urry, D. W.,** Val-Gly-Val-Ala-Pro-Gly, a repeating peptide in elastin, is chemotactic for fibroblasts and monocytes, *J. Cell Biol.,* 99, 870, 1984.
55. **Hascall, V. C. and Hascall, G. K.,** Proteoglycans, in *Cell Biology of Extracellular Matrix,* Hay, E. D., Ed., Plenum Press, New York, 1981, chap. 2.
56. **Hardingham, T., Burditt, L., and Artcliffe, A.,** Studies on synthesis, secretion and assembly of proteoglycan aggregates by chondrocytes, *Prog. Clin. Biol. Res.,* 151, 17, 1984.

57. **Yamada, K. M. and Olden, K.,** Fibronectins — adhesive glycoproteins of cell surface and blood, *Nature (London),* 275, 179, 1978.

58. **Johansson, S. and Hook, M.,** Substrate adhesion of rat hepatocytes: on the mechanism of attachment to fibronectin, *J. Cell Biol.,* 98, 810, 1984.

59. **Pierschbacher, M. D. and Ruoslahti, E.,** Cell attachment activity of fibronectin can be duplicated by small synthetic fragments of the molecule, *Nature (London),* 309, 30, 1984.

60. **Delpech, A. and Delpech, B.,** Expression of hyaluronic acid-binding glycoprotein, hyaluronectin, in developing rat embryo, *Dev. Biol.,* 101, 391, 1984.

61. **Pearlstein, E., Gold, L. I., and Garcia-Pardo, A.,** Fibronectin: a review of its structure and biological activity, *Mol. Cell. Biochem.,* 29, 103, 1980.

62. **Vaheri, A. and Alitalo, K.,** Pericellular matrix glycoproteins in cell differentiation and in malignant transformation, in *Cellular Controls in Differentiation,* Lloyd, C. W. and Rees, D. A., Eds., Academic Press, New York, 1981, 29.

63. **Ruoslahti, E., Pierschbacher, M., Hyman, E. G., and Engvall, E.,** Fibronectin: a molecule with remarkable structural and functional diversity, *Trends Biochem. Sci.,* 7, 188, 1982.

64. **Klebe, R. J. and Mock, P. J.,** Effects of glycosaminoglycans on fibronectin-mediated cell attachment, *J. Cell. Physiol.,* 112, 5, 1982.

65. **Yamada, K. M., Hasegawa, T., Hasegawa, E., Kennedy, D. W., Hirano, H., Hayashi, M., Akiyama, S. K., and Olden, K.,** Fibronectin and interactions at the cell surface, *Prog. Clin. Biol. Res.,* 151, 1, 1984.

66. **Timple, R., Rhode, H., Robey, P. G., Rennard, S. I., Foidart, J. M., and Martin, G. R.,** Laminin — a glycoprotein from basement membranes, *J. Biol. Chem.,* 254, 9933, 1979.

67. **Terranova, V. P., Rao, C. N., Kalebic, T., Margulies, I, M. K., and Liotta, L. A.,** Laminin receptor on breast carcinoma cells, *Proc. Natl. Acad. Sci. U.S.A.,* 80, 444, 1983.

68. **Rao, C. N., Margulies, I. M. K., Tralka, T. S., Terranova, V. P., Madri, J. A., and Liotta, L. A.,** Isolation of a subunit of laminin and its role in molecular structure and tumour cell attachment, *J. Biol. Chem.,* 257, 9740, 1982.

69. **Rao, C. N., Barsky, S. H., Terranova, V. P., and Liotta, L. A.,** Isolation of a tumour cell laminin receptor, *Biochem. Biophys. Res. Commun.,* 111, 804, 1983.

70. **Hewitt, A. T., Varner, H. H., and Martin, G.,** Isolation of chondronectin, in *Immunochemistry of the Extracellular Matrix,* Vol. 1, Furthmayer, H., Ed., CRC Press, Boca Raton, Fla., 1981.

71. **Trelstad, R. L. and Silver, F. H.,** Matrix assembly, in *Cell Biology of Extracellular Matrix,* Hay, E. D., Ed., Plenum Press, New York, 1981, chap. 7.

72. **Ocklind, C., Rubin, K., and Öbrink, B.,** Different cell surface glycoproteins are involved in cell-cell and cell-collagen adhesion of rat hepatocytes, *FEBS Lett.,* 121, 47, 1981.

73. **Wisher, M. H. and Evans, W. H.,** Functional polarity of rat hepatocyte membrane. Isolation and characterization of plasma membrane subfractions from the blood sinusoidal, bile canalicular and contiguous surfaces of the hepatocyte, *Biochem. J.,* 146, 375, 1975.

74. **Cook, J., Hou, E., Hou, Y., Cairo, A., and Doyle, D.,** Establishment of plasma membrane domains. I. Characterization and localization to the bile canaliculus of three antigens externally oriented in the plasma membrane, *J. Cell Biol.,* 97, 1823, 1983.

75. **Skerrow, C. J. and Matoltsy, A. G.,** Isolation of epidermal desmosomes, *J. Cell Biol.,* 63, 515, 1974.

76. **Rees, D. A., Badley, R. A., Lloyd, C. W., Thom, D., and Smith, C. G.,** Glycoproteins in the recognition of substratum by cultured fibroblasts, *Symp. Soc. Exp. Biol.,* 32, 241, 1978.

77. **Staehelin, L. A.,** Structure and function of intercellular junctions, *Int. Rev. Cytol.,* 39, 191, 1974.

78. **Curtis, A. S. G.,** *The Cell Surface: Its Molecular Role in Morphogenesis,* Logos Press, Academic Press, London, 1967, chap. 3.

79. **Doroszewski, J.,** Short term and incomplete cell-substrate adhesion, in *Cell Adhesion and Motility,* Curtis, A. S. G. and Pitts, J. D., Eds., Cambridge University Press, London, 1980, 171.

80. **Gingell, D. and Vince, S.,** Long range forces and adhesion, in *Cell Adhesion and Motility,* Curtis, A. S. G. and Pitts, J. D., Eds., Cambridge University Press, London, 1980, 1.

81. **Rutter, P. R.,** The physical chemistry of the adhesion of bacteria and other cells, in *Cell Adhesion and Motility,* Curtis, A. S. G. and Pitts, J. D., Eds., Cambridge University Press, London, 1980, 103.

82. **Capo, C., Garrouste, F., Benoliel, A. M., Bongrand, P., Ryter, A., and Bell, G.,** Concanavalin A mediated thymocyte agglutination: a model for a quantitative study of cell adhesion, *J. Cell Sci.,* 56, 21, 1982.

83. **Bell, G. I., Dembo, M., and Bongrand, P.,** Cell adhesion. Competition between non-specific repulsion and specific bonding, *Biophys. J.,* 45, 1051, 1984.

84. **Armstrong, P. B.,** On the role of metal cations in cellular adhesion: effect on cell surface charge, *J. Exp. Zool.,* 163, 99, 1966.

85. **Jones, B. M.,** Regulation of the contact behaviour of cells, *Biol. Rev.,* 55, 207, 1980.

86. **Jones, B. M.,** Aspects of cell sorting in aggregates, *Prog. Clin. Biol. Res.,* 151, 275, 1984.

87. **Dembitzer, H. M., Herz, F., Schermer, A., Wolley, R. C., and Koss, L. G.,** Desmosomes development in an *in vitro* model, *J. Cell Biol.*, 85, 695, 1980.
88. **Rassat, J., Robenek, H., and Theman, H.,** Alteration of tight and gap junctions in mouse hepatocytes following administration of colchicine, *Cell Tissue Res.*, 223, 187, 1982.
89. **Tadvalkar, G. and Pinto DaSilva, P.,** *In vitro* rapid assembly of gap junctions is induced by cytoskeletal disrupters, *J. Cell Biol.*, 96, 1279, 1983.
90. **Curtis, A. S. G., Campbell, J., and Shaw, F. M.,** Cell surface lipids and cell adhesion. I. The effects of lysophosphatidyl compounds, phospholipase A_2 and aggregation inhibitory protein, *J. Cell Sci.*, 18, 347, 1975.
91. **Curtis, A. S. G., Chandler, C., and Picton, N.,** Cell surface lipids and adhesion. III. The effects on cell adhesion of changes in plasmalemmal lipids, *J. Cell Sci.*, 18, 375, 1975.
92. **Rutishauser, U., Hoffman, S., and Edelman, G. M.,** Binding properties of a cell adhesion molecule from neural tissue, *Proc. Natl. Acad. Sci. U.S.A.*, 79, 685, 1982.
93. **Edelman, G. M.,** Cell adhesion molecules, *Science*, 219, 450, 1983.
94. **Rutishauser, U.,** Developmental biology of a neural cell adhesion molecule, *Nature (London)*, 310, 989, 1984.
95. **Bushkirk, D. R., Thiery, J-P., Rutishauser, U., and Edelman, G. M.,** Antibodies to neural cell adhesion molecule disrupt histogenesis in cultured chick retina, *Nature (London)*, 285, 488, 1980.
96. **Grumet, M., Rutishauser, U., and Edelman, G. M.,** Neural cell adhesion molecule is on embryonic muscle cells and mediates adhesion to nerve cells *in vitro*, *Nature (London)*, 295, 693, 1982.
97. **Rutishauser, U., Grumet, M., and Edelman, G. M.,** Neural cell adhesion molecule mediates initial interactions between spinal cord neurons and muscle cells in culture, *J. Cell Biol.*, 97, 145, 1983.
98. **Rothbard, J. B., Brachenbury, R., Cunningham, B. A., and Edelman, G. M.,** Differences in the carbohydrate structures of neural cell adhesion molecules from adult and embryonic chick brains, *J. Biol. Chem.*, 257, 11064, 1982.
99. **Edelman, G. M.,** Cell-adhesion molecules: a molecular basis for animal form, *Sci. Am.*, 250, 118, 1984.
100. **Ben-Shaul, Y., Hausman, R. E., and Moscona, A. A.,** Visualization of a cell surface glycoprotein, the retina cognin, on embryonic cells by immuno-latex labeling and scanning electron microscopy, *Dev. Biol.*, 72, 89, 1979.
101. **Rutz, R. and Lilien, J.,** Functional characterization of an adhesive component from embryonic chick neural retina, *J. Cell Sci.*, 36, 323, 1979.
102. **Schubert, D. and LaCorbiere, N.,** Role of a 16S glycoprotein complex in cellular adhesion, *Proc. Natl. Acad. Sci. U.S.A.*, 77, 4137, 1980.
103. **Schubert, D. and LaCorbiere, M.,** Properties of extracellular adhesion mediating particles in myoblast clone and its adhesion-deficient variant, *J. Cell Biol.*, 94, 108, 1982.
104. **Schubert, D., LaCorbiere, M., Klier, F. G., and Birdwell, C.,** A role for adherons in neural retina cell adhesion, *J. Cell Biol.*, 96, 990, 1983.
105. **Damsky, C. H., Richa, J., Solter, D., Knudsen, K., and Buck, C. A.,** Identification and purification of a cell surface glycoprotein mediating intercellular adhesion in embryonic and adult tissues, *Cell*, 34, 455, 1983.
106. **Shirayoshi, Y., Okada, T. S., and Takeichi, M.,** The calcium-dependent cell-cell adhesion system regulates inner cell mass formation and cell surface polarization in early mouse development, *Cell*, 35, 631, 1983.
107. **Burridge, K. and Connell, L.,** A new protein of adhesion plaques and ruffling membranes, *J. Cell Biol.*, 97, 359, 1983.
108. **Enenstein, J. and Furcht, L. T.,** Isolation and characterization of epinectin, a novel adhesion protein for epithelial cells, *J. Cell Biol.*, 99, 464, 1984.
109. **Aplin, J. D., Campbell, S., and Foden, L. J.,** Adhesion of human amnion cells to extracellular matrix. Evidence for multiple mechanisms, *Exp. Cell Res.*, 153, 425, 1984.
110. **Roman, L. M. and Hubbard, A. L.,** A domain-specific marker for the hepatocyte plasma membrane. II. Ultrastructural localization of leucine aminopeptidase to bile canalicular domain of isolated liver plasma membranes, *J. Cell Biol.*, 98, 1488, 1984.
111. **Roman, L. M. and Hubbard, A. L.,** A domain-specific marker for the hepatocyte plasma membrane. III. Isolation of bile canalicular membrane by immunoadsorption, *J. Cell Biol.*, 98, 1497, 1984.
112. **Meier, P. J., Szutl, E. S., Reuben, A., and Boyer, J. L.,** Structural and functional polarity of canalicular and basolateral plasma membrane vesicles isolated in high yield from rat liver, *J. Cell Biol.*, 98, 991, 1984.

Chapter 3

CELL MOTILITY IN DEVELOPMENT

I. CELL LOCOMOTION IN VITRO

A basic mechanism involved in the developmental changes is the motile behavior of the cells. In any attempt to analyze and understand the development of an embryo, a detailed study of this fundamental cell behavior is imperative. In most adult tissues, the cells do not show any spectacular migratory activity; yet when such cells are dissociated from their connections with other cells and intercellular material, they are capable of migration. When an embryonic or adult tissue is explanted on a glass surface as in tissue culture, some individual cells begin to move away. In the culture medium, they do not swim freely; they adhere to the glass or plastic surface and migrate over it.

The motile behavior of single cells in vitro has been studied intensively in attempts to understand cell motility in general. Admittedly, such studies have only a limited value. Cells in the tissues move over other cells or intercellular material and not over inert surfaces like glass or plastic. Besides, most cells are exposed to other tissue components from all sides. In contrast, a cell in vitro is exposed to the glass or plastic on one side and to the culture medium on the "free" surface. Epithelial and endothelial cells when cultured are, however, more akin to their in vivo counterparts. Nonetheless, any cell in vitro is different from its native state. In view of this, the study of cell motility in vitro as a model imposes definite limitations on the information it can yield. In spite of this, it has been a favorite model for experimental studies since it has a number of advantages to offer: access for continuous observation and facility of experimental manipulation. It is hoped that the basic mechanisms of cell motility, albeit modified somewhat, are not likely to change altogether. The wealth of information obtained from in vitro studies justifies this optimism.

It is clear that an integrated approach using the in vitro model as one of the important objects of study will continue to yield valuable information. Innovations in cell culture technology, such as the use of different components of the extracellular material, takes the in vitro condition closer to in vivo, offering at the same time the opportunity to analyze the relative importance of the various factors added. Recent studies on the manner of polymerization of such molecules as collagen, glycosaminoglycans, fibronectin, etc. into organized supramolecular assemblies in vitro enable us to make three-dimensional model matrix structures with controlled physicochemical properties. Refinements in the in vitro technique can be richly rewarding. As an example, we may refer to the work of Erickson and Nuccitelli,[1] who have shown that embryonic quail fibroblast motility can be influenced by small DC electric fields. This is particularly interesting since the magnitude of the effective electric field is small and physiologically obtained in vivo. The relevance of such electric fields within the developing embryos is immediately obvious. How such fields are set up can easily be conjectured. The consequences of the fields can also be inferred. What is most important, however, is that these questions are now in the realm of tangible experimental approach.

A. Cell Surface Perturbations and Cell Motility

Most of the information on cell motility in vitro is derived from the study of fibroblasts. A variety of connective tissue cells are more or less spherical when held in suspension. However, when such cells settle down on a solid substratum, they begin to flatten and assume a characteristic triradiate morphology. Such cells are called fibroblasts. It should be noted, however, that the fibroblast is only a morphological cell phenotype. Fibroblast-like

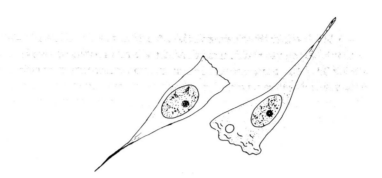

FIGURE 1.   Diagram of moving fibroblasts in vitro. The cells have large lamellipodia in the direction of locomotion. A bleb (clear circular area) is shown on the lamillipodium of the cell on the right. The trailing ends of the moving cells are long and narrow.

cells from embryonic chick cornea, heart, and skin have been shown to be antigenically distinct, at least a part of the difference being surface located.[2] Clearly the term therefore refers to a variety of cells that share some common properties. The fibroblasts can migrate especially away from a group of cells and divide. Locomotion of fibroblasts has been studied by continuous observation, using time lapse cinematography equipped with phase contrast optics.

When a fibroblast moves, the leading edge has a broad, fan-like cytoplasmic lamella closely applied to the substratum. The edge of the lamella undergoes folding movements that beat backward from the edge. This results in a dynamic fold raised from the substratum in the form of a ruffle. Usually when a large ruffled membrane is seen, the cell is moving in that direction (see Figure 1). In addition to the ruffles at the margin, some of them are formed elsewhere on the exposed ("dorsal") side of the fibroblast. Vigorous nonmarginal ruffles have been observed in cells just recovering from experimentally arrested ruffling activity.

Another surface activity of the fibroblastic cells in vitro is the formation of numerous blebs, especially at the margins of the cytoplasm spread over the substratum. The blebs seem to be hemispherical herniations of the cell surface membrane enclosing a thin fluid from the cytoplasm (Figure 1). The ruffles may swell and form blebs. Thus the ruffles and blebs may be related to each other, especially since an intermediate form between them is also observed occasionally. The blebs eventually contract and are withdrawn into the cytoplasm. Blebbing seems to arise from hydrostatic pressure of the cytoplasm, causing herniation of the plasma membrane at a weak point developed, presumably, due to spreading. Addition of sorbitol (which does not diffuse into the cell quickly) raises the external hydrostatic pressure of the medium and causes a collapse of existing blebs. Though ruffling and blebbing as observed in vitro are probably not normal activities of the cell in vivo, a study of the phenomena can indicate the type of membrane changes occurring at the margin of the spreading cell. It also suggests that such activities observed in vitro are modifications of allied cell surface perturbations in vivo.

When fibroblasts in culture encounter small particles (colloidal gold or nickel, carbon, etc.), the latter are frequently lifted up and carried over the cell surface in a highly consistent pattern.[3] Eventually such particles are carried rearward, accumulated around the nucleus or the trailing end of the cell. The particles are not endocytosed as inferred from the observation that sometimes they fall off at the cell margin or stick to neighboring cells. Definitive evidence has been obtained by ultrastructural studies to show that these particles are carried over the cell surface and not in the interior of the cell.[4] The particle transport that has been observed as a general feature of many types of cells in vitro is always accompanied by cell

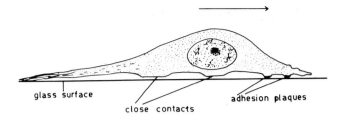

FIGURE 2.   Schematized diagram of a fibroblast on glass surface, shown in vertical section. The arrow indicates the direction of locomotion of the cell.

locomotion. The precise mechanism by which a particle is carried over the plasma membrane is not understood clearly. It may be due to the rearward flow of the membrane material that is continually reassembled at the leading edge of the cell. Alternatively, it could be a movement of the plasma membrane quite like a wrinkle in a carpet that can be moved from one place to another without displacing the carpet material itself. Whatever the mechanism, the phenomenon indicates that a variety of cell surface modulations can be brought about by a particle coming in contact with the cell.

B. Movement of a Fibroblast

When a fibroblast is placed on a substratum, it changes from a spherical shape into a flat one. The cell spreads in such a manner as to conform to the contours of the substratum. It is probable that this is a generalized behavior of cells related to other activities such as phagocytosis.[5] As the cell flattens, it assumes a thin disc shape with a somewhat irregular outline. At the thickest part where the nucleus is located, the cell may be 2 to 3 μm, but at the periphery it is very thin, 0.1 to 0.2 μm. Eventually, ruffling activity starts in the lamella. If there are several lamellipodia in different directions, there is no displacement of the cell. On the other hand, if one of them predominates, the cell becomes polarized, with the lamella constituting the leading edge. The cell has now the typical "fibroblast" shape. Just behind the leading edge of the front pseudopodium (dominant lamellipodium), the cell surface makes close contacts with the substratum. These are called focal contacts or adhesion plaques. They are oriented areas of contact with the substratum. Generally they are 1 μm long and 0.25 μm wide, with their long axis pointing in the direction of locomotion (Figure 2). As the lamella progresses, new focal contacts are established. Besides the focal contacts, there are somewhat broader areas of approach of the plasma membrane to the substratum. These are called close contacts, but in fact they are not as close as the adhesion plaques. Whereas the close contacts do not leave behind any mark when detached from the substratum, the adhesion plaques leave behind some material.

When fibroblasts move in vitro, they do so jerkily, apparently by breaking adhesions of the cell surface at the trailing end and establishing new ones with the substratum at the leading edge. The thin trailing end is observed to move intermittently by sudden detachment, and its cytoplasm is resorbed into the main cell body (Figure 3). The early observations of Abercrombie et al.[3] and Ambrose[6] on the behavior of fibroblasts in vitro have been confirmed and extended by many subsequent workers. Ambrose[6] suggested that the chief locomotory organ of the migrating fibroblast is the leading lamellipodium. He also suggested that there may be a submembranous contractile system responsible for the locomotion. Subsequent studies have largely vindicated these suggestions. The importance of ruffles in locomotion is, however, doubtful. It has been shown that the leading edge can advance without ruffling. In fact, rapidly moving cells ruffle least frequently. The free (leading) edge of the lamellipodium shows regional and temporary protrusions and retractions. At one point across its

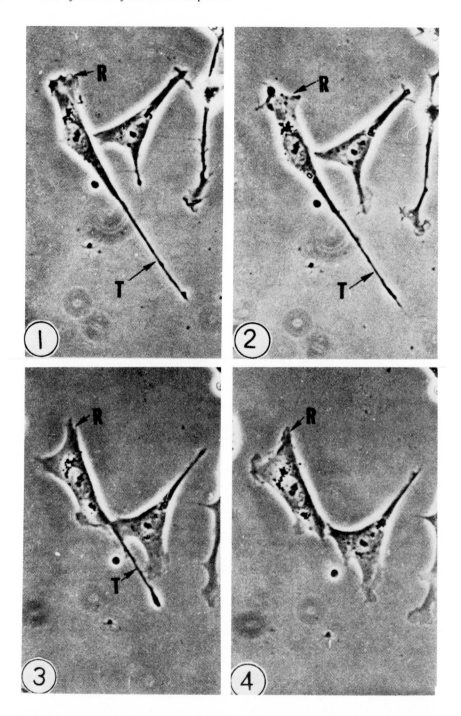

FIGURE 3.   A series of phase contrast micrographs of a rat embryo fibroblast, taken over a period of 40 min, showing the movement of the cell body and resorption of the tail region (T). The movement is in the direction of the ruffled leading edge (R), toward the top left of the photographs. The other cell in the pictures may be used as a stationary reference mark. (From Goldman, R. D., Schloss, J. A., and Strager, J. M., in *Cell Motility,* Book A, Goldman, R., Pollard, T., and Rosenbaum, J., Eds., Cold Spring Harbor Laboratory, New York, 1976, 240. With permission.)

FIGURE 4.   Diagram of an actin filament.

breadth, the lamellipodium might be advancing while the nearby region is retracting. However, the net result of this is the advancement of the cell since the protruding activity is more frequent and lasting compared with retraction. It is this dynamic protrusion-retraction activity that takes the cell along, and the role of the ruffles seems to be insignificant.

C. The Role of Cytoskeletal Apparatus in Cell Motility

The contractile apparatus of the fibroblastic cells has been shown to consist of microfilament bundles found in the stretched part of the cell. Living cells observed by phase contrast show the presence of "stress fibers" in the trailing part of the cell. In electron micrographs, the stress fibers are represented by microfilament bundles. The contractile apparatus of the fibroblast seems to function in a manner similar to muscle as shown by the contraction of the tail of a glycerinated model of fibroblast treated with exogenous ATP in the presence of Mg^{2+}. (The glycerinated model consists of cells that are extracted with 50% glycerol in saturated salt solutions for 1 to 2 hr). This phenomenon is strikingly similar to contraction of normal living fibroblasts and seems to depend on actomyosin contraction. Rounded (detached) cells contain no microfilament bundles, but do contain a meshwork of structures which have been shown to be made of actin. When such cells are allowed to spread and attach to the substratum, the microfilament bundles are assembled again, even in the absence of protein synthesis. A precursor-product relationship between the meshwork and microfilament bundles has been shown to be likely by the demonstration that both take up rabbit skeletal muscle heavy meromyosin and appear fuzzy ("decorated") in electron micrographs. This indicates that the meshwork and filamentous structure contain actin. Reorganization of actin filaments in living fibroblasts has recently been studied by Wang.[7]

Considerable attention has been paid to the role of actin and the associated proteins in cell locomotion. Actin is evolutionarily highly conserved in amino acid sequence,[8] which suggests the strong selection pressure to maintain the structure and interaction sites for similar functional requirements. The fibrous actin is polymerized from the globular actin monomers in the cytoplasm. The monomeric and polymerized actin (G-actin and F-actin, respectively) are in equilibrium in the living cell. The polymerized fibrous form occurs as a twin-stranded helical structure (Figure 4; for details see De Rosier and Tilney[9]). Binding of heavy meromyosin on the actin filaments can be demonstrated by electron microscopy. Fluorescently labeled proteins can be used as probes to detect cytoskeletal structures in cell models.[10] The "decorated" fibers have a characteristic barbed appearance, with the "barbs" pointing in one direction and the pointed ends in the opposite. On the basis of this, the fibrous actin filament has two morphologically distinct ends, the "barbed" and the "pointed" (Figure 5).

In the past 5 years or so, considerable information has been obtained on the formation of microfilaments and their organization into stress fibers. The actin polymerization process has been elucidated by several groups of workers, particularly Oosawa and Asakura[11] and Wegner.[12] The first step in the polymerization of G-actin is its activation. This process is dependent on K^+ and renders the monomers more resistant to proteolysis. "Nucleation" is the next step. It appears that in order to initiate the formation of fibrous polymers, a trimer, which is thought to be the first stable beginning of fiber elongation, has to be present. Since the formation of trimers is very low (it depends on the third power of the monomer concentration), it is an important rate-limiting factor in the process. Elongation is the further

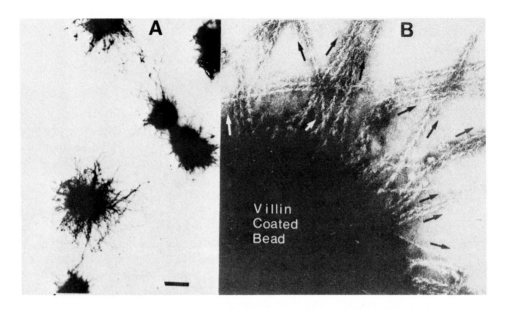

FIGURE 5.   Polymerization of actin on villin coated polystyrene beads. The beads were coated with villin core and added with salt to a solution of actin to allow polymerization from the surface of the bead. The S1 fragment was added and an aliquot was applied to a carbon coated electron microscopic grid and stained with 1% uranyl acetate. (A) Low power magnification (bar represents 1 μm) showing the numerous filaments attached to each bead. (B) Higher magnification showing the polarity of attached filaments. Note the uniform polarity of the filaments in which the arrowheads point away from the beads. Arrows indicate the polarity of nearby filaments. (From Weber, K. and Glenney, J. R., Jr., *Cold Spring Harbor Symp. Quant. Biol.*, 46, 541, 1982. With permission.)

step. Monomers are added to the ends of the "nucleus" (trimer) or longer filaments. This is a reversible process, and hence the monomers already added can also dissociate. An interesting aspect of actin polymerization is that chain elongation occurs at different rates at the barbed and pointed ends. Elongation at the barbed end is more rapid than at the other end. It has been estimated that ≈70 monomers are added and two dissociated every second at the barbed end; at the other end about 20 are added and one is lost during the same time interval. Consequently the filament grows by about 90 monomers per second. During or after polymerization, the ATP bound to the monomer is split to ADP. Since a monomer with either bound ATP or ADP can be added to the growing filament, there are two different association rate constants. Fiber elongation is diagrammatically represented in Figure 6. It will be seen that there are four distinct association rate constants (two for the addition at the barbed or pointed end, each again for ATP or ADP association). Similarly there are four distinct dissociation constants. The net growth of a filament is therefore controlled by the concentrations of different molecules and their aggregates. (For further details and references, see Pollard and Craig.[13]) Obviously this is a complex process involving several possible controls. As though this was not enough, there are additional complications due to the presence of other actin binding proteins that modify the polymerization process.

Recently a number of actin-binding proteins have been identified in various organisms.[14] This has paved the way to understanding some aspects of cell locomotion. When combined with a complexing protein, profilin, the actin monomers are unable to polymerize, and thus they can be sequestered from the pool. By cross-linking with other proteins, actin filaments can form isotropic gels or bundles. Some cross-linking proteins have recently been described. These are filamin, α-actinin, and others. The conditions required for gelation with these proteins are not identical. Filamin promotes side-to-side as well as end-to-side interactions of actin filaments and increases the rate of actin polymerization. Filamin-cross-linked gelation

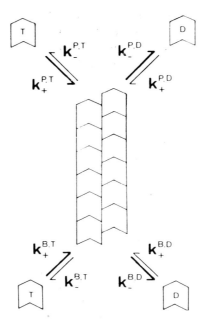

FIGURE 6. Kinetic model of actin polymerization. The barbed (B) and pointed (P) ends of the filaments and monomers are indicated by the shapes of the subunits; k_+, association rate constant; k_-, dissociation rate constant. T denotes bound ATP and D denotes bound ADP. (Reproduced from *The Journal of Cell Biology*, 1981, 88, 654, by copyright permission of the Rockefeller University Press.)

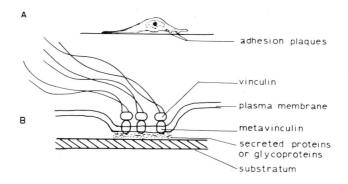

FIGURE 7. Diagram to illustrate the probable arrangement of different proteins at an adhesion plaque. (A) A fibroblast spread on glass surface showing the position of plaques. (B) One of the adhesion plaques drawn to show further details. The long curved lines directed to the left represent actin filaments.

of actin seems to be involved in the function of membrane ruffles and stress fibers. Unlike the filamin which is a long, flexible molecule, α-actinin is a rod-like actin-cross-linking protein. In addition to its well-known occurrence in the Z-disks of skeletal muscle, in the nonmuscle cells it has been demonstrated in association with stress fibers and focal contacts. It has also been suggested that α-actinin dissociates actin-profilin complexes, thus enriching the actin monomer pool to be utilized in polymerization. Vinculin is another protein that cross-links actin filaments. It occurs at the ends of stress fibers where they terminate at the focal contacts of lamellipodia.[15-17] It is not an integral membrane protein. Metavinculin, an integral membrane protein, acts as an anchor for vinculin (Figure 7).[18] Another actin-binding protein, fimbrin, has been demonstrated in the intestinal brush border microvilli. It also occurs in the membrane ruffles of cultured cells. Fimbrin binds F-actin at a saturation level of one fimbrin per two to three actin monomers, thus forming tightly packed bundles. Villin, yet another actin-binding protein occurring in the microvilli, also cross-links with actin filaments forming bundles.

Finally, one more protein has to be mentioned in this context. Recent studies have shown that tropomyosin exists in several isoforms, synthesized within single cells. The isoforms differ in their molecular weights. Tropomyosins interact with and modify the action of the various actin-binding proteins.[19] It has been suggested that this could be yet another factor modulating the organization of microfilaments in living cells.

An elaborate mechanism regulating the elongation of actin filaments, their cross-linking and depolymerization seems to exist in cells. The so-called capping proteins interfere with the growth of actin filaments by blocking one of its ends. The majority of such proteins described so far cap the barbed end of the growing actin filament, thus inhibiting the addition of new monomers. When introduced into living tissue culture cells by microinjection, they can disrupt the existing microfilament bundles.[20] This action affects the focal contacts. Thus the action of the capping proteins resembles the effect of cytochalasin B, a fungal alkaloid. Disruption of microfilament bundles by capping proteins or cytochalasin occurs only in living cells; in detergent-treated "models" there is no disruption of the existing microfilament bundles. The difference is significant in indicating the mode of action of these agents. Disruption occurs in living cells in which the ends of microfilaments are in a dynamic equilibrium with free actin monomers. In dead cells there are no biochemical activities of association/dissociation, and so the agents are ineffective. These facts also indicate that the capping proteins do not sever the microfilaments.

There are other actin-binding proteins, however, which can cleave or fragment the actin filaments. These include fragmin, gelsolin, and others (see Table 1). An important factor involved in this complex interplay of proteins is the role of Ca^{2+}. Some of the actin-binding proteins are Ca^{2+} sensitive, whereas the others are not. Figure 8 illustrates diagrammatically the interplay of some of these proteins. It is not yet clear how these various proteins interact and bring about any identifiable action in cell motility. Obviously, the cross-linking proteins, which organize the actin filaments into bundles or isotropic gels, could be the agents bringing about gel/sol properties of the cytoplasm. A classical theory of amoeboid movement[21] contends that gelation and contraction of the cytoplasm at the trailing end of an amoeba pushes forward the less viscous "plasmasol." At the progressing end, the plasmasol moves to the surface and gelates as ectoplasm and flows backwards. Though this theory does not account for all the facts known at present, it suggests that the interconversion of gel-sol could be an important activity in living cells, and the actin cross-linking proteins obviously have a definite role in this. Besides, however, the role of the other actin binding proteins has to be recognized and integrated into a more comprehensive theory of cell locomotion. No such attempt has been made so far. The recently identified actin-binding proteins have been mentioned here first to indicate the complexity of the mechanism that regulates cellular deformations and locomotion, and second, to point out that future work aimed at elucidating the role of these factors will be richly rewarding. For a comprehensive review on actin binding proteins, see Weeds.[22]

Do the stress fibers have any role in the locomotion of fibroblasts in vivo? Are the stress fibers indeed associated with locomotory activity or are they formed as a response to the artificial substratum (glass or plastic) on which the cells move? Recent work on the locomotion of fibroblasts has provided some additional information on the significance of the stress fibers. Herman et al.[23] have presented evidence that the stress fibers are associated with fibroblasts, which have slowed down their locomotory activity. In fact, no stress fibers are found in the fibroblasts initially growing out from a tissue explant and migrating rapidly.

Table 1
DISTRIBUTION OF ACTIN-BINDING PROTEINS

Class	Families	Subunit (mol wt)	Protozoa	Other lower eukaryotes	Vertebrates
Bind actin monomers	Profilin	12—15K	+	+	+
	Depactin/actophorin	16—20K	+	+	+
Bind end of actin filaments	Capping protein	29K + 31K	+	+	+
	Fragmin/severin	40—45K	0	+	+
	Accumentin	65K	0	0	+
	Gelsolin/villin	90—95K	0	0	+
Bind along actin filaments	Tropomyosin	30—40K	0	0	+
Cross-link actin filaments to each other	Gelactins	23—38K	+	0	0
	Fascin/fimbrin	55—70K	0	+	+
	α-Actinin	90—100K	+	+	+
	Actin-binding protein/ filamin	250K	0	0	+
Cross-link actin filaments to other structures	Brush border 110K	110K	0	0	+
	Spectrin	220—260K	+	+	+
	Microtubule-associated protein 2	260K	0	0	+
Myosins	Myosin/myosin⁻ⁱⁱ	175—220K	+	+	+
	Myosin⁻ⁱ	125—130K	+	0	0

Note: Actin-binding proteins are grouped into classes by their established properties and subdivided into families according to physical properties and presumed mechanisms of action. This classification is arbitrary and will probably be modified as more data become available. +, The protein has been identified in these cells; 0, the protein has not yet been identified in these cells (or rarely, for example in the case of tropomyosin in Protozoa, the protein has been sought, but not found). Mol wt, molecular weight.

Reprinted by permission from *Nature*, 312, 403. Copyright© 1984, Macmillan Journals Limited.

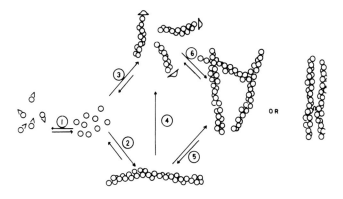

FIGURE 8. Diagram to illustrate actin polymerization and its control. 1, Free actin monomers in equilibrium with the monomers bound depolymerizing protein; 2, monomers in equilibrium with polymerized long filaments; 3, monomers in equilibrium with short polymerized filaments capped by capping proteins; 4, breaking of long filaments into shorter fragments; 5 and 6, short or long filaments cross-linked to form meshwork or bundles.

Thus, although the stress fibers are functionally contractile organelles, their relation with locomotion of cells does not seem to be simple. It has been suggested by Burridge[24] that the stress fibers develop in response to unphysiological conditions of culture in which the cells adhere to plastic more strongly than in vivo. The large actomyosin filament bundles (i.e., the stress fibers) presumably arise as a result of the cell pulling off a point of firm

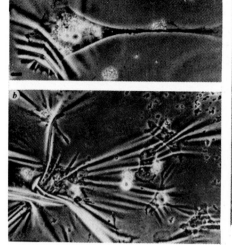
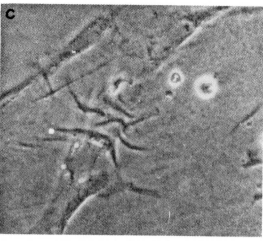

FIGURE 9.   Tissue culture cells spreading on a thin layer of silicone rubber. (*a*) An individual 3T3 cell which has produced numerous folds in the rubber support beneath the leading margin. (*b*) Complex pattern of distortion produced in the supporting rubber sheet, by a group of chick heart fibroblasts. (*c*) Distortion of a collagen substratum by cellular traction producing a pattern similar to the wrinkling of silicone rubber. Scale bars: *a*, 10 μm; *b*, 100 μm; *c*, 50 μm. (Reprinted by permission from *Nature*, 290, 249. Copyright © 1981. Macmillan Journals Limited.)

adhesion. The ordering of the microfilaments into bundles could be (at least partly) a physical consequence of strong adhesion and the tension generated on the substratum. Studying the movement of cells over flexible silicone rubber and collagen film substrates, Harris and colleagues[25,27] have shown that highly motile cells do not deform the flexible substrate, but more sedentary (more strongly adhesive?) fibroblasts produce streaks at right angles to the direction of locomotion (Figure 9). They also suggest that the traction could have a morphogenetic role in ordering the eventual orientation of collagen matrix from a meshwork in vivo.

D. Interactions Among Moving Cells

When a moving fibroblast encounters another cell in the culture, the ruffling activity is paralyzed and the cell stops locomotion. This tendency of fibroblasts to stop locomotion on coming in contact with another cell is called contact inhibition of locomotion.[28] The phenomenon of contact inhibition of locomotion has been considered as the cause of monolayer formation in culture, i.e., contact inhibition of growth. it is also known that certain fibrosarcoma cells do not show contact inhibition of locomotion and growth so that they can move over one another and also continue to divide. A consequence of this is the overlapping of the cells in culture. These observations are of fundamental importance in understanding the social behavior of cells in vitro as well as in vivo.

Locomotion of a large variety of cells has been studied in vitro, and it now appears that contact inhibition of locomotion is not the cause of contact inhibition of growth (i.e., cell division). Besides, the mechanism by which locomotion of colliding cells is inhibited is not uniform for all types of cells. This has necessitated the recognition of two types of contact inhibition of locomotion. When the leading lamellae of some cells such as chick heart fibroblasts meet during locomotion, the ruffling activity in both the colliding cells is paralyzed almost immediately. The leading edges separate by retraction. When the leading lamella of such a fibroblast meets the side of another fibroblast, the lamella of the first may continue to move under the other fibroblast for a short while, but eventually it stops. The most immediate result of the contact between the cells is thus the stopping of the ruffling activity.

This type of contact inhibition of locomotion is recognized as "Type 1".[29] The "Type 2" contact inhibition is one in which there is no inhibition of the ruffling activity though the forward locomotion is inhibited. It has been suggested that this is due to the cell-substrate adhesiveness being greater than the cell adhesion. Thus the failure of locomotion is due to the failure of cell-cell contacts being established. In other words, Type 2 inhibition is substrate dependent.

It must be remembered that the in vitro behavior of cells is essentially artificial or a more or less modified version of the in vivo behavior. In recent years, it is becoming increasingly clear that the intercellular material plays an important role in the control and integration of cell behavior in vivo. Collagen, the major extracellular constituent of most tissues, is known to influence the adhesive and migratory behavior of cells. It is common in tissue culture practice to coat the glass surface with collagen (or denatured collagen, i.e., gelatin) so as to obtain proper adhesion and improved cell growth. Stenn et al.[30] have shown that migrating epidermis cells require continued collagen synthesis for movement. Schor[31] studied the behavior of a variety of cells on the surface of plastic petri dishes, collagen film, and fibrous matrix. Besides, the behavior of the cells within a three-dimensional matrix of collagen was also studied.[32,33] Though some cells showed an initial lag in growth, they reached more or less similar final densities while growing in the different environments. Their locomotory behavior was, however, markedly different. HeLa cells when plated on the plastic surface formed tightly packed colonies of polygonal cells indicating that they did not migrate away from the center of growth. On the other hand, on the surface of collagen film or matrix, the cells were separated by gaps greater than several cell diameters. Within the collagen matrix, the cells grew into compact clusters.

An interesting difference between the behavior of epithelial and fibroblastic (and other nonepithelial) cells was also revealed in the study. Epithelial cells do not infiltrate into the three-dimensional matrix, whereas the other cells do. Infiltration of cells into collagen matrix does not involve collagenase activity.[30] Fibroblasts seem to be stimulated to migrate by the three connective tissue collagen types as reported by Postlethwaite et al.,[34] who observed that collagenase degradation products and even small peptides resulting from collagen digestion can stimulate migration. There is also evidence to show that cell motility is influenced by fibronectin.[35] The foregoing account emphasizes the importance of the extracellular substances in regulating the locomotion of cells in embryos. Conversely, the positional stability of cells in adult tissues and its loss during metastasis could also depend on their interactions with the extracellular material. In subsequent chapters, we shall have occasion to return to this topic.

II. THE NEURAL CREST

The vertebrate neural crest is unique in the diversity of embryonic sites to which it migrates and the variety of the cellular phenotypes it gives rise to. Originating from the well defined neural folds (the lateral ridges of the neural plate), the neural crest cells are found between the neural tube and the dorsolateral epidermis (Figure 10). From this location, they migrate over long distances and differentiate. The versatility in the differentiation of the neural crest cells is remarkable (see Table 2; Figure 11). Several important questions arise from a study of the development of the neural crest. What causes the cells to leave their initial location and migrate to other embryonic sites? Do the cells have a predetermined mechanism that leads them to their definitive locations or does the environment guide them? Are the neural crest cells determined, i.e., is their differentiating capacity fixed or is it subject to modification according to the location they reach? These and related questions have constituted the subject of extensive embryological research for a long time. Currently, with the advent of new and refined experimental techniques, these questions are beginning to get answers.

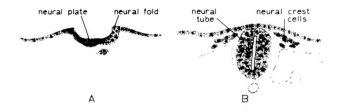

FIGURE 10.   Diagrammatic transverse sections of vertebrate embryos to show the origin of the neural crest.

Table 2
MAJOR DERIVATIVES OF THE AVIAN NEURAL CREST

1. Neural derivatives[a]
    Sensory ganglia (including neurons, glia, sheath, and satellite cells)
        Trigeminal[p] (V)
        Root[p] (VII)
        Superior (IX)
        Jugulare (X)
        Spinal dorsal root ganglia
    Sympathetic ganglia (nonvascular components only) and tissues
        Paravertebral (chain)
        Prevertebral (coeliac, mesenteric, adrenal, retro-aortic complexes)
        Adrenal medulla
    Parasympathetic ganglia (nonvascular components only)
        Ciliary
        Submandibular, ethomid, sphenopalatine, otic, lingual
        Intrinsic visceral
        Meissner's, Auerbach's, Remak's, pelvic plexes
    Schwann sheath cells
    Supporting cells (glia, satellite cells), but not neurons
        Geniculate (VII) and acoustic (VIII) ganglia[pp]
        Petrosal ganglion[pp] (XI)
        Nodose ganglion[pp] (X)
2. Cartilages, bones, muscles, and connective tissues (from cranial neural crest only)[b]
    Visceral arch cartilages*
        1st Arch: Meckel's quadrate
        2nd Arch: columella, stylohyal, basihyal (entoglossal)
        Branchial arches: all remnants of basi-, cerato-, and epi-branchials
    Loose connective tissue of face, tongue, and lower jaw
3. Other neural crest derivatives[b]
    Pigment cells
        Melanocytes of dermis, mesenteries, internal organs, epidermis, etc.
        Melanophores of iris
    Secretory cells
        Carotid body Type I cells
        Calcitonin-producing cells of ultimobranchial body
    Chondrocranial cartilages*
        Ethmoid, interorbital septum
        Anterior and posterior orbitals
        Sclera, nasal capsule
    Lower jaw bones
        Dentary, angular, supra-angular, opercular (splenial)
        Articular, quadratojugal
    Upper jaw, palatal, and cranial vault bones'
    Maxilla, premaxilla, palatine, nasal, prefrontal, lacrimal, frontal (rostral part)
        Anterior paraspheneoid (rostorum), pterygoid, squamosum (squamous temporal)
        Scleral ossicles

Table 2 (continued)
MAJOR DERIVATIVES OF THE AVIAN NEURAL CREST

Muscles
    Ciliary muscles (striated)
    Some cranial vascular and dermal smooth muscles (see below)
    Visceral arch-derived muscles (minor component)
Corneal endothelium and stromal fibroblasts
Mesenchymal component of adenohypophlysis, lingual gland, parathyroid, thymus, thyroid
Dermis and subcutaneous adipose of face, jaw, and upper neck, and mesenchyme adjacent to oral epithelium
Leptomeninx of the diencephalon and telencephalon
Arterial wall smooth muscle and elastic fiber tissues, but NOT endothelial layer derivatives of visceral arch
    vessels, including parts of internal, external, and common carotids, 4th arch component of systemic aorta,
    pulmonary (6th arch) vessels
Venous wall components, excluding endothelium in superficial facial, oral and jaw regions

[a]   Some (p) or all (pp) of these neurons are of epidermal placode origin.
[b]   Structures listed (*) are those found in the embryo either prior to replacement by bone or (') before fusion of
    ossified skeletal elements.

From Noden, D. M., *Specificity of Embryological Interactions*, Garrod, D. R., Ed., Chapman & Hall, London,
1978, chap. 1. With permission.

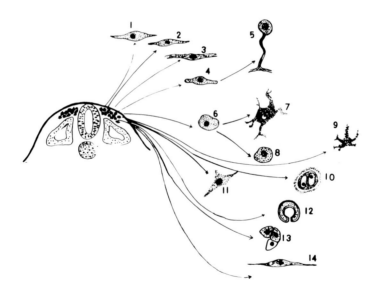

FIGURE 11.   Some of the cell types derived from the neural crest. 1, Pial cell;
2, arachnoid cell; 3, neurilemmal cell; 4, bipolar neuroblast; 5, cell of the spinal
ganglion; 6, sympathetic neuroblast; 7, sympathetic ganglion cell; 11, head mesenchyme cell; 12, Schwann sheath cell; 13, secretory cell of adrenal medulla; 14,
arterial wall smooth muscle cell.

In order to gain an insight into the subject, a detailed study of the development of the neural
crest is essential. An outline of this is presented below and may be supplemented by other
literature.[36-39]

A. Embryology
    The following description refers largely to the development of the avian neural crest. It
should not be concluded, however, that information is not available regarding other vertebrates. References to the literature may be found in the reviews on the subject.[36-39]

The precise pattern of the migration of cells within a developing embryo cannot be discerned from a simple histological study in sections since the cells are often not characterized by any morphological or other distinguishing feature. Nicholas[40] has described a histochemical method of staining the neural crest cells of mouse embryos. The feasibility of applying this technique in experimental studies remains to be proven. For unmistakable identification of the cells, the use of an appropriate marker is therefore imperative. In experiments involving grafting anatomically defined regions of neural crest to replace corresponding or different regions of a host embryo, the migration of cells can be followed provided that the grafted cells have incorporated a radioactive isotope such as ³H-thymidine. Autoradiography of the sections of such grafted embryos at different intervals from the time of grafting has yielded considerable information.[41,42] The method has, however, an important drawback: the label is continuously "diluted" as the cells undergo repeated divisions and hence detection of the labeled cells offers problems.

Recently the chick-quail chimaeras have been proved to be an elegant alternative to radioactive labeling. The Japanese quail (*Coturnix coturnix Japonica*) nuclei have a characteristically large heterochromatin associated with the nucleolus, which can be stained by the Feulgen-Rosenbeck or other procedures. Individual cells of the quail tissue can therefore be identified in tissue sections. There is no danger of "dilution" of the marker, no matter how many times the cells divide. Employing the technique of grafting quail neural crest in chick host embryos, considerably accurate information has been obtained by Le Douarin and colleagues.[38,39] More recently, antibodies raised against neural crest cell surface antigens have been used for detecting the cells in histological sections. The method consists of raising antibodies in mice against chick neural crest cells (e.g., ciliary ganglia). Sections of chick embryos are exposed to these antibodies, which bind the neural crest cells specifically. Bound antibodies are revealed by FITC-conjugated antimouse subclass immunoglobulin.[43] Besides these techniques, scanning electron microscopy has permitted the revelation of considerable details of the patterns of neural crest cell migration.[44-48]

The neural crest develops from a transitory structure, the neural fold, as shown in Figure 9. Separation of the neural crest from the neural primordium and the epidermis is correlated with the folding of the neural plate into a tube in the pre-otic region. In the post-otic and trunk region, the wedge shaped neural crest is formed along with or a little earlier than the formation of the somites. In the chick embryo, differentiation proceeds in an antero-posterior direction, and thus there is a gradient of maturing of the neural crest from the anterior to the posterior end.[44]

In the region of the forebrain, the neural crest cells migrate around the optic stalks, Some crest cells remain dorsal to the forebrain and eyes. Some of these cells move between the lens and superficial ectoderm, giving rise to the corneal endothelial and stromal layers and contributing to the iris. Other crest cells of this region are associated with the maxillary process during facial morphogenesis. In the midbrain region, the migrating crest cells follow a ventro-lateral course contributing to the maxillary process and some mesenchyme. Most of the crest cells of the hind-brain region move ventrally between the epidermis and mesoderm. These become subdivided segmentally and aggregate adjacent to the loci of trigeminal and facial nerve emergence. With a contribution from the ectodermal placodes, they form cranial ganglia in this region. The other crest cells also move ventrally and get subdivided segmentally, giving rise to the mandibular and hyoid arches (Figure 12).

In the trunk region, the crest cells begin to migrate as an unsegmented mesenchyme ventro-laterally towards the dorsal margin of the somite. The pathways taken by these cells depend on their relation to the adjacent somite: (1) the cells that face the middle of the somite accumulate between the neural tube and the apex of the somite and will give rise to the dorsal root ganglion; (2) the cells facing the intersomitic space rapidly migrate towards the para-aortic region to form the primary sympathetic chain; and (3) at somitic levels,

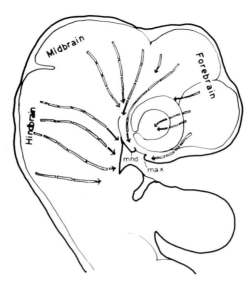

FIGURE 12. Diagram of the head region of a chick embryo showing the routes of migrating neural crest cells; max, maxilla; mnd, mandible.

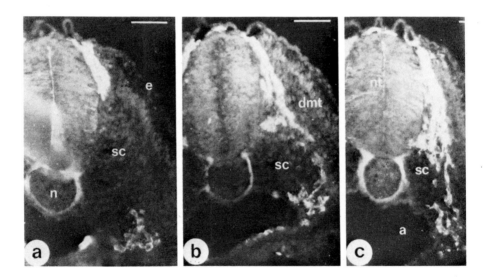

FIGURE 13. Neural crest cells of a chick embryo labeled by a fluorescent monoclonal antibody. The three photographs represent level of the middle of a trunk somite (a), level intermediate between mid- and inter-somite levels (b), and level close to the intersomitic space (c). a, Aorta; dmt, dermomyotome; e, ectoderm; n, notochord; nt, neural tube; sc, sclerotome. Scale bar 50 μm. (From Vincent, M. and Thiery, J-P., *Dev. Biol.*, 103, 468, 1984. With permission.)

intermediate between the two, many cells localize between the neural tube and somite. Some of these cells later pass between the dermamyotome and sclerotome and settle near the aorta where they contribute to the autonomic structures. Later in development, some crest cells migrate, remaining close to the surface ectoderm, and give rise to the pigment cell precursors after being associated with the dermis (Figure 13). This detailed description of the pathways of the crest cells was possible by the use of antibodies as specific labels.[43]

From the region of somites 1 to 7, the crest cells migrating ventrally colonize the dorsal mesentery and give rise to the enteric ganglia. Crest cells from the region of somites 7 to

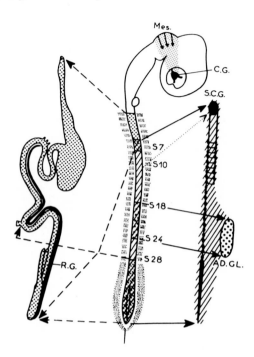

FIGURE 14.   Levels of origin of adrenomedullary cells and autonomic ganglion cells. The spinal neural crest caudal to the level of the 5th somite gives rise to the ganglia of the orthosympathetic chain. The adrenomedullary cells originate from the spinal neural crest between the levels of somites 18 and 24. The vagal neural crest (somites 1 to 7) gives rise to the enteric ganglia of the preumbilical region, the ganglia of the postumbilical gut originating from both the vagal and lumbosacral neural crest (posterior to the somite-28 level). The ciliary ganglion (C.G.) is derived from the mesencephalic crest (Mes.). AD. GL. = adrenal gland; S.C.G. = superior cervical ganglion. (From Le Douarin, N. M., Teillet, M. A., and Le Lièvere, C., in *Cell and Tissue Interactions,* Lash, J. W. and Burger, M. M., Eds., Raven Press, New York, 1977, 11. With permission.)

28 give rise to neurons of the sensory and sympathetic chain of ganglia, aortic, and adrenal plexuses. Within this region, the crest cells of somites 18 to 24 level give rise to the adrenal medulla. Behind the level of somite 28, the crest derivatives give rise to enteric as well as sympathetic ganglion cells. This complex pattern of migration and differentiation (Figure 14) was revealed by Le Douarin and associates from elaborate experiments in which small segments of the neural groove, along with the neural folds, were obtained from quail donor embryos and grafted in chick embryos, replacing the corresponding host structures. Subsequently the migration of the crest cells was observed making use of the quail nuclear marker.[38,39]

An important aspect of the biochemical differentiation of the two major neuronal systems needs to be considered in this connection. They differ in the neurotransmitter substances produced in them. The enteric neurons are characterized by the synthesis of acetyl choline, whereas the other neural crest derivatives (viz., the sympathetic ganglia and adrenomedullary cells), are characterized by catecholamines (adrenaline and noradrenaline; see Figure 15). Obviously the biochemical pathways leading to the phenotypic expression of the cells are different. An important question in this connection is: are the neural crest cells distributed along the cranio-caudal axis as predetermined phenotypes so that the highly patterned and precise distribution of different cell types is due to their initial location? Alternatively, if it is not so, does the environment of their migratory route and the definitive location determine their differentiation? Heteroplastic grafting experiments were done to answer these questions. Quail neural crest cells of the adrenomedullary region grafted at the vagal level in chick embryos of similar age colonize the gut and give rise to functional cholinergic enteric ganglia.

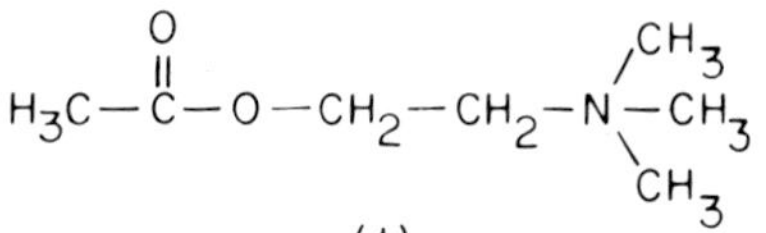

$$H_3C-\overset{\overset{\displaystyle O}{\parallel}}{C}-O-CH_2-CH_2-\overset{\overset{\displaystyle CH_3}{\diagup}}{\underset{\underset{\displaystyle CH_3}{\diagdown}}{N}}-CH_3$$

(1)

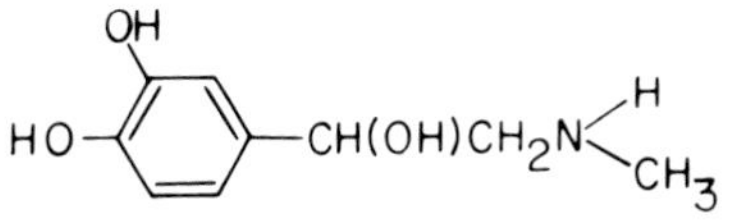

$$HO-\text{(benzene ring with OH)}-CH(OH)CH_2N\overset{\overset{\displaystyle H}{\diagup}}{\underset{\underset{\displaystyle CH_3}{}}{}}$$

(2)

$$HO-\text{(benzene ring with OH)}-CH(OH)CH_2NH_2$$

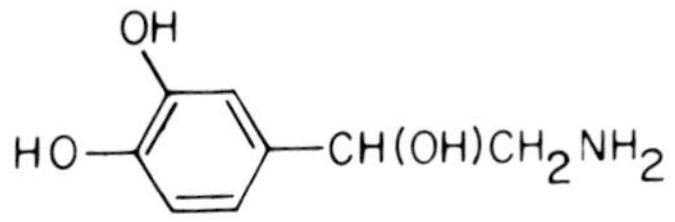

(3)

FIGURE 15. Chemical formulas of acetyl choline (1), adrenaline (2), and noradrenaline (3).

Neural crest from the mesencephalic or rhombencephalic region of quail grafted in chick at the adreno-medullary level populated the adrenal medulla and differentiated into adrenergic cells.

From these experimental results, it could be concluded that the migratory route available to the crest cells and/or their definitive location determine the pathway of their differentiation. In other words, the potential to differentiate into cholinergic or adrenergic phenotypes exists in the entire crest.[38] Other experiments by Le Douarin and colleagues have shown that the differentiation of the crest neuronal cells is influenced by the nonneuronal cells in the environment, and the differentiated state develops as a result of "inducing" influences of the environment. Even more striking is the observation that the neurotransmitter phenotype is labile and reversible. When cholinergic ciliary ganglia from 4.5- to 6-day quail embryos were implanted into 2-day chick embryos at the level of somites 18 to 24, the grafted cells began to disperse in the host trunk region. The dispersal was far from random: the quail cells participated in the formation of various host crest derivatives (viz., sympathetic ganglia, adrenomedullary cords, and others), which have adrenergic cells. Significantly, the quail cells were virtually never found in the host sensory ganglia. The quail cells were shown to contain catecholamines. Thus the cells of the ciliary ganglia, which normally are never adrenergic at any stage, could change their pathway of differentiation as dictated by their new environment.[49]

The heterotopic transplantation experiments described above have shown that the premigratory neural crest cells are pleuripotential. However, the possibility exists that the population of these premigratory cells is heterogeneous, with latent phenotypic differences. Monoclonal antibodies raised against different crest derivatives have indicated such a heterogeneity.

B. The Environment of Moving Crest Cells

A somewhat detailed account of the development of the neural crest has been given here because it offers the opportunity to investigate virtually all the aspects of developmental biology of current interest. During their emigration, the neural crest cells follow definite routes of migration and interact with the environment en route as well as at the destination. As a consequence of these interactions, the cells differentiate into distinct phenotypes. In this process, the inherent nature of the cells is, of course, important: for example, noncrest cells do not develop into crest derivatives even if they are placed in the appropriate environment. In this sense, the cells may be considered as already "determined" or in the process of differentiation. However, certain major events of differentiation may be more or less determined by the environment. Heterotopic grafting experiments have shown that in general, the premigration crest cells are not determined and develop more or less according to the environment in which they differentiate rather than with reference to the location of their origin in the embryo. However, there are instances of some previous determination also. Cephalic crest cells implanted at the trunk level contribute to the peripheral ganglia of the host embryo but, in addition, give rise to some visceral skeletal structures which is their "fate" with reference to the origin. Diversification of the crest derivatives is thus a complex process and seems to take place at different points of time and probably in several steps.

Many general features of embryonic development are strikingly exhibited by the neural crest. Elegant experimental techniques are now available, thus facilitating a detailed examination of this remarkable system in many different ways. Our main purpose of dealing with the neural crest is to understand the role of the cell surface in cellular migration and differentiation. Several investigators have shown that some extracellular material is organized in the form of a fibrillar matrix between the dorsal epidermis and the neural tube of amphibian embryos. The fibrillar material has been shown to be collagen associated with glycosaminoglycans.[50] Similar fibrillar matrix has been demonstrated in the developing chick embryos. These observations have been made in studies using histochemical and ultrastructural examination of the embryos.[51] In particular, the scanning electron micrographs[44,46] show the fibrillar matrix and its relation to the cells in the region of the embryo where migration of the crest cells begins.

An interesting feature of the fibrillar matrix is the dorso-ventral orientation of the collagen fibers as demonstrated by Löfberg et al.[52] in Axolotl embryos. At the sites of contact with the fibrillar matrix, the neural crest cells begin to migrate.[52] From this, it has been suggested that the extracellular material offers contact guidance to the migrating cells. The migrating neural crest cells have been shown to exhibit flattening and polarization in their morphology similar to fibroblasts. On the contrary, nonmigrating neural crest cells are rounded.[44] The crest cells show considerable overlapping and seem to possess no contact-inhibition property. An interesting mutant in the Mexican Axolotl (*Ambystoma mexicanum*) is the "white" which, as a homozygous recessive, is characterized by the failure of pigment cells or their crest precursors to migrate laterally into the subepidermal space. Grafting a piece of "dark" (wild type) epidermis permits the migration of the pigment cells. Conversely, grafting patches of "white" epidermis on "dark" hosts prevents the migration of melanoblasts underneath it. It appears that the mutant has some local defect in the epidermis. Spieth and Keller[53,54] have described the morphology of the crest cells and their extracellular matrix in this interesting mutant. Similar studies on the genetic control of pigment cell pattern have been reported by Hallet and Ferrand.[55]

The occurrence of hyaluronate in association with the basement membrane of the epidermis and neural tube has been demonstrated by several workers.[56,57] This material hydrates and expands, thus probably opening up tissue spaces and facilitating cell migration. Hyaluronic acid synthesis seems to be the first necessary event, forming a space ahead of neural crest cell migration pathway. The work of Thiery and associates[58,59] indicates that the presence of fibronectin is also associated with the pathways of neural crest cell migration. Greenberg et al.[60] suggested that fibronectin is both chemotactic and chemokinetic to the crest cells. It is possible that the organization of the fibrillar material with hydrated hyaluronate may itself act as a stimulant for the beginning of orientation and migration of the crest cells. Recent work has clarified some uncertainties in this matter. Erickson et al.[47] have shown that neural crest cells grafted in the middle of the presumptive pathway of migration move both backward and forward, thereby indicating that the importance of chemotactic stimulus is much less than suggested on the basis of conjecture. High cell density has been recognized as another factor influencing crest cell migration.[61] Whether this can exert a vectorial influence is, however, not clear.

Bronner and Cohen[62] developed a technique of introducing cultured quail melanocytes in host chick embryos to trace their pathway. When introduced in the early embryos' neural crest region, the melanocytes migrate ventrally, past the sensory and sympathetic ganglionic sites, eventually localizing in their definitive location near the aortic plexus, adrenal gland, metanephric mesenchyme, or gut. It is to be noted that these cells are nonmotile in vivo or in vitro. Their migratory activity is induced by the environmental factors. Bronner-Fraser and Cohen[63] injected melanocytes freshly isolated from the skin of 9- to 11-day quail embryos. These cells also exhibited a similar migratory activity. Many noncrest cells, however, do not migrate in this way.[47] From these observations, it may be concluded that the nature of the cells is also important in responding to the environment in which they are placed. An additional observation of Erickson et al.[47] is that Sarcoma-180 cells show extensive migratory abilities along the neural crest routes. It appears that some inherent motile behavior (expressed or latent) in the cells is essential for their locomotion, and the environment offers some guiding influence. A novel technique of using latex beads as probes of neural crest pathway has recently been described by Bronner-Fraser.[64] This provides a powerful tool to ascertain the relative importance of inherent motile behavior and the extracellular matrix in defining the pattern of migration. The arrest of the cells at their definitive location could be due to the development of some physical barrier ahead of the front of migration, besides the absence of material to guide and orientate. Concomitant loss of fibronectin and hyaluronic acid at the definitive sites seems to be a significant correlation in this context.

Artificial three-dimensional matrices consisting of the various components of the extracellular matrix permit a detailed ''in vitro'' study of the behavior of neural crest cells in an environment more akin to in vivo than conventional in vitro culture or planar substrata coated with various materials. An additional advantage of the system is that the matrix components can be tested individually or in any desired experimental combinations. Tucker and Erickson[65] have reported elaborate investigations on the morphology and behavior of quail neural crest cells in such artificial matrices. Native collagen at optimum concentration in the matrix permits rapid locomotion of the crest cells. Denatured collagen does not have this property. Many cells move in the collagen matrix with a rounded trailing edge (cf. locomotion of fibroblasts in vitro). Also, some bipolar cells are found. Collagen matrix supplemented with fetal calf serum and chick embryo extract enhances crest cell migration considerably. Addition of native fibronectin to the collagen matrix is not much effective in enhancing cell migration. However, a fragment of fibronectin with intact cell-adhesion and collagen-adhesion sites, added to the collagen matrix, increases cell migration to the same extent as do fetal calf serum and embryo extract. Hyaluronic acid added to collagen is not effective at low doses; at higher doses it is inhibitory. Chondroitin sulfate reduces cell locomotion.

These observations of Tucker and Erickson[65] suggest that collagen stimulates cell migration. Fibronectin also interacts with other matrix components and may be stimulatory to crest cell migration. Hyaluronic acid is, however, not stimulatory; it can aid by opening up spaces due to hydration. The precise manner in which the matrix components interact and guide the crest cells is not yet clear. What is abundantly clear, however, is that the matrix components modulate the behavior of the crest cells and bring about their patterned migration — initiation, progress, and stopping.

Probably the matrix components act in unison and modulate cell adhesiveness. The neural cell adhesion molecules (N-CAMs) discussed in the previous chapter have also been reported to have some role in the migration of the neural crest cells.[58,59] When the neural crest cells first appear, they have the N-CAMs at their surface. The matrix around them is rich in fibronectin. When they divide and migrate away, the N-CAMs on their surface diminish. At their destination, the N-CAMs reappear in the surrounding spaces. This seems to facilitate their clumping to form structures such as ganglia.[66,67] These observations cannot be built into a comprehensive theory on the role of the N-CAMs in the development of the different crest derivatives. Nevertheless, further attempts at elucidating the interrelations of the adhesion molecules and the extracellular matrix that exhibits a very wide variety in different embryonic locations will undoubtedly be rewarding.

Interesting and important as these recent studies are, it seems to be premature to suggest any generalized mechanism for the complex migratory behavior of the neural crest cells. Some important questions have yet to receive clear answers. For example, what precise anatomical details determine the different routes taken by the crest cells of the head and trunk is not clear. Besides the extracellular matrix, does the cellular environment have any role in guiding the crest cells? As mentioned earlier, cell density could be a decisive factor. It is not known if the cellular environment exerts any other influence. Further, little is known about the factors that lead the cells to choose between cholinergic and adrenergic pathways. However more precise information is likely to be forthcoming in the near future.

An important question, which ought to receive more attention, is pertaining to the early events in the origin of the crest cells. The dorsal ectoderm of the early neurula has three distinct parts. In the median (supranotochordal) region is the presumptive neural plate that thickens and develops into the central nervous system. The neural plate is flanked by the neural folds, the precursors of the neural crest. The cells constituting the neural folds lose their epithelial character and migrate like mesenchymal cells. Lateral to the neural folds is the presumptive epidermis. The cells constituting this layer remain thin and in fact may become even thinner.

The mechanism causing the diverse behavior of the three regions has been conserved during vertebrate evolution. In particular, how the cells organized as an epithelium constituting the neural folds, lose their morphology, and acquire mesenchymal characteristics is not clear. Greenberg and Hay[68] have reported that a variety of epithelial cells held within a three-dimensional collagen matrix can assume fibroblastic shape and migrate like connective tissue cells. Schor,[31] on the other hand, has shown that epithelia just placed on the surface of three-dimensional matrices cannot penetrate and move in. Initially the neural crest cells are not in contact with any matrix other than their basement membrane. A precisely localized mechanism to destabilize the basement membrane must therefore exist. Clearly then, the influence of extracellular matrix in initiating the development of the neural crest does not seem to be a satisfactory explanation. Though the later events of migration of the cells may be facilitated or even initiated by the extracellular matrix, the very early event, i.e., the loss of epithelial character seems to be owing to some other distinct mechanism that still remains a mystery. Another question, which may be posed in this connection, is whether the closure of the neural tube is an essential prerequisite for the timely release of the crest cells. Studying chick embryos treated with hyaluronidase before neural tube closure, Anderson and Meier[69] conclude that it is not so.

III. THE PRIMORDIAL GERM CELLS

A striking example of single cells migrating in the embryo and reaching their definitive position far away from the place of origin is found in the primordial germ cells of the vertebrates. The extra gonadal origin of the primordial germ cells has been demonstrated in birds,[70] mammals,[71-74] and amphibians.[75] In the posterior half of the vertebrate embryo, the rudiments of gonads appear as thickened longitudinal strips of mesoderm lateral to the dorsal mesentery (Figure 16). The thickenings are called genital or germinal ridges. It is now established that sperm and ova are descendants of the primordial germ cells that migrate to

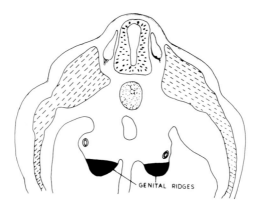

FIGURE 16.   Diagrammatic transverse section of a vertebrate embryo
showing the genital ridges.

the genital ridges. An interesting feature of the migration of the primordial germ cells is that they are not carried to their destination by general morphogenetic movements such as those characteristic of gastrulation. They move as individual cells and thus resemble the neural crest cells. Here again we can pose several questions regarding their migratory behavior. Are the cells attracted by the tissue which is their destination? Or, alternatively, do the cells move randomly and not leave the gonads once having reached there? Are there any preferred paths of their migration?

The fact that primordial germ cells of vertebrates have an extra-gonadal origin has now been established. In the amphibian, *Xenopus*, the presumptive primordial germ cells have been detected at the vegetal pole in the earliest cleavage stage. During the neurula and tail bud stages, they are found deeply located within the endodermal cell mass below the archenteron. In birds, they have been located in a horseshoe shaped area anterior to the head process, at the border of the pellucid and opaque areas, associated with the lower layer of the blastoderm.[76] In mammals (mouse, man, etc.), they are found associated with the yolk sac[57] (Figure 17). Detection of the primordial germ cells has been possible because they can be stained rather specifically by certain histochemical methods. Mintz and Russel[78] demonstrated the cells in mouse embryos by staining for the enzyme alkaline phosphatase. In the chick, they are rich in glycogen content and are stained by the Schiffe reagent after periodic acid oxidation.[79-81] The chick primordial germ cells are also characterized by their large size (10 to 15 μm across, compared with other tissue cells that are generally 6 to 10 μm).

Classical experimental studies have shown that the cells stained by these special methods are indeed primordial germ cells. Mintz and Russel[78] demonstrated that embryos of a mutant strain of sterile mouse have few primordial germ cells stainable for alkaline phosphatase, whereas the normal embryos have many such cells. In birds, extirpation of the area of the blastoderm containing these cells results in sterile embryos. More interesting are the results of Simon,[82] who obtained chick embryos in parabiotic pairs, one with, and the other without, the primordial germ cells. It was found that the embryo, which was operated to remove the horseshoe shaped area eliminating the primordial germ cells, develops gonads with normal germ cells in case it has a common circulation with the parabiotic partner. This clearly indicates that the germ cells migrate through blood circulation.

The origin of the presumptive primordial germ cells seems to be endodermal.[83,84] However, Eyal-Giladi et al.[85] have shown that the avian primordial germ cells have an epiblastic origin. They employed the technique of grafting quail epiblast of prestreak blastoderms in chick blastoderms of similar age. Reciprocal grafts were also made. Primordial germ cells, detected

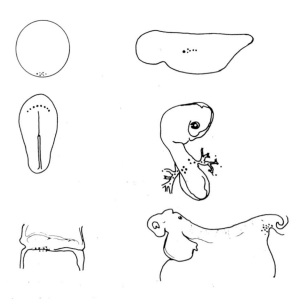

FIGURE 17. Diagrams of amphibian (upper), avian (middle), and mammalian (lower) embryos. The primordial germ cells are shown by dark dots in an early (left) and a later (right) stage of development.

later, showed that they arise from the epiblast of the donor species. According to these workers, this raises an important question on the homology of the primordial germ cells. In a subsequent chapter, we shall discuss the evidence to show that the chick embryonic endoderm arises from the early epiblast. In fact, from the point of view of comparative embryology, the avian epiblast includes all the three presumptive germ layers of the embryo. Since the avian epiblast normally gives rise to the definitive embryonic endoderm, the observations of Eyal-Giladi et al.[85] can still be reconciled with the generally held view that the primordial germ cells are of endodermal origin. In any case, their observations open an important question that needs to be examined in higher vertebrate embryos where distinct hypoblast and endoblast exist.

The pathway followed by the primordial germ cells in reaching the genital ridges seems to be different in the various groups of vertebrates investigated. In the amphibia, they appear to pass through a cellular environment, wedging their way through the dorsal mesentery and finally reaching the genital ridges. The role of glycoconjugates of the extracellular matrix in the migration of the cells has been demonstrated.[86] In mammals, this also seems to be so. In birds, however, as already indicated, the evidence shows that they take a vascular course in their migration. Reynaud[87] introduced turkey primordial germ cells intravascularly into host chick embryos that were exposed to UV in the germ cell region. It was found that the sterile genital ridges were populated by turkey germ cells. Delayed introduction of the turkey cells indicates that the competence of gonads to retain the germ cells decreases with the age of the host embryo, disappearing completely after the 5th day of incubation. Similar experiments have been reported using the quail marker.[88]

IV. CONCLUDING REMARKS

One could readily think of a number of questions regarding cell migration in a variety of situations. Migration of leukocytes to the sites of injury or inflammation is not merely of academic interest. The knowledge of this phenomenon is of immense practical value. Information on this subject may be obtained by reference to Hayashi et al.[89] and Keller et al.[90] There is little question that research on motility as a general attribute of cells is of relevance to all workers in the vast field of biomedical science. Besides the neural crest and primordial germ cells, many examples of cell migration in embryonic development could be thought of. However, the discussion in this chapter is sufficient to highlight the best known aspects of the problem. In subsequent chapters, we shall again discuss cell motility in connection with normal and morbid developmental processes.

REFERENCES

1. **Erickson, C. A. and Nuccitelli, R.,** Embryonic fibroblast motility and orientation can be influenced by physiological electric fields, *J. Cell Biol.,* 98, 296, 1984.
2. **Garrett, D. M. and Conrad, G. W.,** Fibroblast-like cells from embryonic chick cornea, heart and skin are antigenically distinct, *Dev. Biol.,* 70, 50, 1979.
3. **Abercrombie, M., Heaysman, J. E. M., and Pegrum, S.,** The locomotion of fibroblasts in culture. II. 'Ruffling', *Exp. Cell Res.,* 60, 437, 1970.
4. **Harris, A. K.,** Cell surface movements related to cell locomotion, in *Locomotion of Tissue Cells,* Ciba Foundation Symp., Vol. 14 (new series), Associated Publishers, Amsterdam, 1973, 5.
5. **Grinnel, F.,** Fibroblast spreading and phagocytosis: similar cell responses to different sized substrata, *J. Cell. Physiol.,* 119, 58, 1984.
6. **Ambrose, E. J.,** The movement of fibrocytes, *Exp. Cell Res.,* 8(Suppl.), 54, 1961.
7. **Wang, Y. L.,** Reorganization of actin filaments in living fibroblasts, *J. Cell Biol.,* 99, 1478, 1984.
8. **Vandekerckhove, J. and Weber, K.,** Amino acid sequence of *Physarum* actin, *Nature (London),* 276, 720, 1978.
9. **De Rosier, D. J. and Tilney, L. G.,** How actin filaments pack into bundles, *Cold Spring Harbor Symp. Quant. Biol.,* 46, 525, 1982.
10. **Sanger, J. W., Mittal, B., and Sanger, J. M.,** Interaction of fluorescently labelled contractile proteins with the cytoskeletons in cell models, *J. Cell Biol.,* 99, 918, 1984.
11. **Oosawa, S. and Asakura, S.,** *Thermodynamics of the Polymerization of Proteins,* Academic Press, London, 1975.
12. **Wegner, A.,** Head to tail polymerization of actin, *J. Molec. Biol.,* 108, 139, 1976.
13. **Pollard, T. D. and Craig, S. W.,** Mechanism of actin polymerization, *Trends Biochem. Sci.,* 7, 55, 1982.
14. **Pollard, T. D.,** Actin binding protein evolution, *Nature (London),* 312, 403, 1984.
15. **Geiger, B.,** A 130K protein from chicken gizzard: its localization at the termini of microfilament bundles in cultured chicken cells, *Cell,* 18, 193, 1979.
16. **Burridge, K. and Feramisco, J. R.,** Microinjection and localization of a 130K protein in living fibroblasts: a relationship to actin and fibronectin, *Cell,* 19, 587, 1980.
17. **Burridge, K.,** Studies on α-actinin and vinculin: proteins of the adhesion plaque, *Anat. Rec.,* 1(Suppl.), 51, 1983.
18. **Siliciano, J. D. and Craig, S. W.,** Metavinculin — a vinculin-related protein with solubility properties of a membrane protein, *Nature (London),* 300, 533, 1982.
19. **Payne, M. R. and Rudnick, S. E.,** Tropomyosin as a modulator of microfilaments, *Trends Biochem. Sci.,* 9, 361, 1984.
20. **Füchtbauer, A., Jockusch, B. M., Maruta, H., Kiliman, M. W., and Isenberg, G.,** Disruption of microfilament organization after injection of capping proteins into living tissue culture cells, *Nature (London),* 304, 361, 1983.
21. **Goldacre, R. J.,** The role of cell membrane in the locomotion of amoebae and the source of the motive force and its control by feedback, *Exp. Cell Res.,* 8(Suppl.), 1, 1961.
22. **Weeds, A.,** Actin binding proteins — regulators of cell architecture and motility, *Nature (London),* 296, 811, 1982.
23. **Herman, I. M., Crisona, N. J., and Pollard, T. D.,** Relation between cell activity and distribution of cytoplasmic actin and myosin, *J. Cell Biol.,* 90, 84, 1981.
24. **Burridge, K.,** Are stress fibres contractile?, *Nature (London),* 294, 69, 1981.
25. **Harris, A. K., Stopak, D., and Wild, P.,** Fibroblast traction as a mechanism for collagen morphogenesis, *Nature (London),* 290, 249, 1981.
26. **Stopak, D. and Harris, A. K.,** Connective tissue morphogenesis by fibroblast traction, *Dev. Biol.,* 90, 383, 1982.
27. **Harris, A. K., Jr.,** Tissue culture cells on deformable substrata: biomechanical implications, *J. Biochem. Eng.,* 106, 19, 1984.

28. **Abercrombie, M.,** Contact-dependent behaviour of normal cells and the possible significance of surface changes in virus-induced transformation, *Cold Spring Harbor Symp. Quant. Biol.,* 27, 427, 1962.
29. **Heaysman, J. E. M.,** Contact inhibition of locomotion: a reappraisal, *Int. Rev. Cytol.,* 55, 49, 1978.
30. **Stenn, K. G., Madri, J. A., and Roll, F. J.,** Migrating epidermis produces AB₂ collagen and requires continued collagen synthesis for movement, *Nature (London),* 277, 229, 1979.
31. **Schor, S. L.,** Cell proliferation and migration on collagen substrata *in vitro, J. Cell Sci.,* 41, 159, 1980.
32. **Schor, S. L., Allen, T. D., and Harrison, C. J.,** Cell migration through three-dimensional gels of native collagen fibres: collagenolytic activity is not required for the migration of two permanent cell lines, *J. Cell Sci.,* 46, 171, 1980.
33. **Schor, S. L., Schor, A. M., and Bazilla, G. W.,** The effects of fibronectin on the migration of human foreskin fibroblasts and syrian hamster melanoma cells into three-dimensional gels of native collagen fibres, *J. Cell Sci.,* 48, 301, 1981.
34. **Postlethwaite, A. E., Seyer, J. M., and Kang, A. H.,** Chemotactic attraction of human fibroblasts to type I, II, and III collagens and collagen-derived peptides, *Proc. Natl. Acad. Sci. U.S.A.,* 75, 871, 1978.
35. **Ali, I. U. and Hynes, R. O.,** Effects of LETS glycoprotein on cell motility, *Cell,* 14, 439, 1978.
36. **Weston, J. A.,** The migration and differentiation of neural crest cells, *Adv. Morphogenesis,* 8, 41, 1970.
37. **Noden, D. M.,** Interactions directing the migration and cytodifferentiation of avian neural crest cells, in *Specificity of Embryological Interactions,* Garrod, D. R., Ed., Chapman & Hall, London, 1978, chap. 1.
38. **Le Douarin, N. M.,** Migration and differentiation of neural crest cells, *Curr. Top. Dev. Biol.,* 16, 32, 1980.
39. **Le Douarin, N. M.,** *The Neural Crest,* Cambridge University Press, London, 1982.
40. **Nichols, D. H.,** Neural crest formation in the head of the mouse embryo as observed using a new histological technique, *J. Embryol. Exp. Morphol.,* 64, 105, 1981.
41. **Weston, J. A.,** A radioautographic analysis of the migration and localization of trunk neural crest cells in the chick, *Dev. Biol.,* 6, 279, 1963.
42. **Noden, D. M.,** An analysis of the migratory behaviour of avian cephalic neural crest cells, *Dev. Biol.,* 42, 106, 1975.
43. **Vincent, M. and Thiery, J-P.,** A cell surface marker for neural crest and placodal cells: further evolution of peripheral and central nervous system, *Dev. Biol.,* 103, 468, 1984.
44. **Bancroft, M. and Bellairs, R.,** The neural crest cells of the trunk region of the chick embryo studied by SEM and TEM, *Zoon,* 4, 73, 1976.
45. **Tosney, K. W.,** The early migration of neural crest cells in the trunk region of the avian embryo. An electron microscopic study, *Dev. Biol.,* 62, 317, 1978.
46. **Tosney, K. W.,** The segregation and early migration of cranial neural crest cells in the avian embryo, *Dev. Biol.,* 89, 13, 1982.
47. **Erickson, C. A., Tosney, K. W., and Weston, J. A.,** Analysis of migratory behaviour of neural crest and fibroblastic cells in embryonic tissues, *Dev. Biol.,* 77, 142, 1980.
48. **Erickson, C. A. and Weston, J. A.,** An SEM analysis of neural crest migration in the mouse embryo, *J. Embryol. Exp. Morphol.,* 74, 97, 1983.
49. **Le Douarin, N. M. and Smith, J.,** Differentiation of avian autonomic ganglia, in *Autonomic Ganglia,* Elfvin, L-G., Ed., John Wiley & Sons, New York, 1983, 427.
50. **Corsin, J.,** Lé materiel extracellulaire au cours du dévelopment du chondrocrâne des amphibiens: mise en place et constitution, *J. Embryol. Exp. Morphol.,* 38, 139, 1977.
51. **Hay, E. D.,** Fine structure of embryonic matrices and their relation to cell surface in ruthenium red-fixed tissues, *Growth,* 42, 399, 1978.
52. **Löfberg, J., Ahlfors, K., and Fällström, C.,** Neural crest cell migration in relation to extracellular matrix organization in the embryonic Axolotl trunk, *Dev. Biol.,* 75, 148, 1980.
53. **Spieth, J. and Keller, R. E.,** Neural crest cell behaviour in white and dark larvae of *Ambystoma mexicanum:* differences in cell morphology, arrangement and extracellular matrix as related to migration, *J. Exp. Zool.,* 229, 91, 1984.
54. **Keller, R. E. and Spieth, J.,** Neural crest cell behaviour in white and dark larvae of *Ambystoma mexicanum:* time-lapse cinematographic analysis of pigment cell movement *in vivo* and in culture, *J. Exp. Zool.,* 229, 109, 1984.
55. **Hallet, M-M. and Ferrand, R.,** Quail melanoblast migration in two breeds of fowl and in their hybrids: evidence for a dominant genic control of the mesodermal pigment cell pattern through the tissue movement, *J. Exp. Zool.,* 230, 227, 1984.
56. **Manasek, F. J. and Cohen, A. M.,** Anionic glycopeptides and glycosaminoglycans synthesized by embryonic neural tube and neural crest, *Proc. Natl. Acad. Sci. U.S.A.,* 74, 1057, 1977.
57. **Greenberg, J. H. and Pratt, R. M.,** Glycosaminoglycan and glycoprotein synthesis by cranial neural crest cells *in vitro, Cell Differentiation,* 6, 119, 1977.
58. **Thiery, J-P., Duband, J. L., and Delouvée, A.,** Pathways and mechanisms of avian trunk neural crest cell migration and localization, *Dev. Biol.,* 93, 324, 1982.

59. **Duband, J. L. and Thiery, J-P.,** Distribution of fibronectin in the early phase of avian cephalic neural crest cell migration, *Dev. Biol.,* 93, 308, 1982.
60. **Greenberg, J. H., Seppä, S., Seppä, H., and Hewitt, A. T.,** Role of collagen and fibronectin in neural crest cell adhesion and migration, *Dev. Biol.,* 87, 259, 1981.
61. **Rovasio, R. A., Delouvée, A., Yamada, K. M., Timpl, R., and Thiery, J-P.,** Neural crest cell migration: requirements for exogenous fibronectin and high cell density, *J. Cell Biol.,* 96, 462, 1983.
62. **Bronner, M. E. and Cohen, A. M.,** Migratory patterns of cloned neural crest melanocytes injected into host chicken embryos, *Proc. Natl. Acad. Sci. U.S.A.,* 76, 1843, 1979.
63. **Bronner-Fraser, M. and Cohen, A. M.,** Analysis of neural crest ventral pathway using injected tracer cells, *Dev. Biol.,* 77, 130, 1980.
64. **Bronner-Fraser, M.,** Latex beads as probes of neural crest pathway: effects of laminin, collagen and surface charge on bead translocation, *J. Cell Biol.,* 98, 1947, 1984.
65. **Tucker, R. P. and Erickson, C. A.,** Morphology and behaviour of neural crest cells in artificial three-dimensional extracellular matrices, *Dev. Biol.,* 104, 390, 1984.
66. **Edelman, G. M.,** Cell adhesion molecules, *Science,* 219, 450, 1983.
67. **Edelman, G. M.,** Cell adhesion and morphogenesis: the regulator hypothesis, *Proc. Natl. Acad. Sci. U.S.A.,* 81, 1460, 1984.
68. **Greenburg, G. and Hay, E. D.,** Epithelia suspended in collagen gels can lose polarity and express characteristics of migrating mesenchymal cells, *J. Cell Biol.,* 95, 333, 1982.
69. **Anderson, C. B. and Meier, S.,** Effect of hyaluronidase treatment on the distribution of cranial crest cells in the chick embryo, *J. Exp. Zool.,* 221, 329, 1982.
70. **Swift, C. H.,** Origin and early history of the primordial germ cells in the chick, *Am. J. Anat.,* 15, 483, 1914.
71. **Brambell, F. W. R.,** The development and morphology of the gonads of the mouse. I. The morphogenesis of the indifferent gonad and the ovary, *Proc. R. Soc. London Ser. B,* 101, 391, 1927.
72. **Everett, N. B.,** Observational and experimental evidences relating to the origin and differentiation of the definitive germ cells in mice, *J. Exp. Zool.,* 92, 49, 1943.
73. **Witschi, E.,** Migration of the germ cells of human embryos from the yolk sac to the primitive gonadal folds, *Carnegie Inst. Wash. Contrib. Embryol.,* 32, 67, 1948.
74. **Mintz, B.,** Continuity of the female germ cell line from embryos to adult, *Arch. Anat. Microsc. Morphol. Exp.,* 48, 155, 1959.
75. **Smith, L. D.,** The role of a "germinal plasm" in the formation of primordial germ cells in *Rana pipiens, Dev. Biol.,* 14, 330, 1966.
76. **Clawson, R. C. and Domm, L. V.,** Origin and early migration of primordial germ cells in the chick: a study of the stages definitive primitive streak through 8 somites, *Am. J. Anat.,* 125, 87, 1969.
77. **Mintz, B.,** Germ cell origin and history in the mouse: genetic and histochemical evidence, *Anat. Rec.,* 127, 335, 1957.
78. **Mintz, B. and Russel, E. S.,** Gene-induced embryological modifications of primordial germ cells in the mouse, *J. Exp. Zool.,* 134, 207, 1957.
79. **Fujimoto, T., Ninomiya, T., and Ukeshima, A.,** Observations of the primordial germ cells in blood samples from the chick embryo, *Dev. Biol.,* 49, 278, 1976.
80. **Mc Kay, D. G., Hertig, A. T., Adams, A. C., and Danziger, S.,** Histochemical observation on the germ cells of human embryos, *Anat. Rec.,* 117, 201, 1953.
81. **Simon, D.,** Contribution a l'étude de la circulation et du transport des gonocytes primaires dans les blastodermes d'oiseaux cultivées *in vitro, Arch. Anat. Microsc. Morphol. Exp.,* 49, 93, 1960.
82. **Simon, D.,** Association de blastodermes d'oiseaux en culture *in vitro.* Application de cette méthode a la migration des gonocytes primaires d'un embryon à un autre embryon, *Colloq. Int. Centre Nat. Rech. Sci.,* 101, 269, 1961.
83. **Nieuwkoop, P. D. and Sutasurya, L. A.,** *Primordial Germ Cells in the Chordates,* Cambridge University Press, London, 1979.
84. **Michael, P.,** Are the primordial germ cells (PGCs) in Urodela formed by the inductive action of the vegetative yolk mass?, *Dev. Biol.,* 103, 109, 1984.
85. **Eyal-Giladi, H., Ginsburg, M., and Farbarov, A.,** Avian primordial germ cells are epiblastic in origin, *J. Embryol. Exp. Morphol.,* 65, 139, 1981.
86. **Delbos, M., Gipouloux, J-D., and Saidi, N.,** The role of glycoconjugates in the migration of anuran amphibian germ cells, *J. Embryol. Exp. Morphol.,* 82, 119, 1984.
87. **Reynaud, G.,** Transfert de cellules germinales primordiales de dindon à l'embryon de poulet par injection intravasculaire, *J. Embryol. Exp. Morphol.,* 21, 485, 1969.
88. **Reynaud, G.,** Etude de la localisation des cellules germinales primordiales chez l'embryon de caille japonaise au moyen d'une technique d'irradiation aux rayons ultraviolets, *C.R. Acad. Sci. Paris,* 282, 1195, 1976.

89. **Hayashi, H., Honda, M., Shimokawa, Y., and Hirashima, M.,** Chemotactic factors associated with leucocyte emigration in immune tissue injury, their separation, characterization and functional specificity, *Int. Rev. Cytol.,* 89, 179, 1984.
90. **Keller, H. U., Naef, A., and Zimmerman, A.,** Effects of colchicine, vinblastine and nocodazole on polarity, motility, chemotaxis and cAMP levels of human polymorphonuclear leucocytes, *Exp. Cell Res.,* 153, 173, 1984.

Chapter 4

THE CELLULAR SLIME MOLDS

I. SIMPLE ORGANISMS AS MODELS OF MORPHOGENESIS

Understanding the basic mechanisms underlying the development of complex animal embryos is not easy for several reasons. First, the fertilized egg is already a highly specialized cell with its own structural and chemical organization. Its complexity increases rapidly even before the establishment of the germ layers has occurred. The complex patterns of migration exhibited by the cells during gastrulation suggest that the cells have already differentiated to a certain extent. The differentiation is not revealed by the synthesis of any easily identifiable product characterizing these cells. Nevertheless these cells are already endowed with the ability to perform morphogenetic movements. The migration exhibited by the different cells is far from identical. Each cell *in situ* has a characteristic direction and extent of migrating program that is executed with a remarkable precision and in cooperation with the other cells. This state of "differentiation" is more or less labile since the cells removed from their in vivo association with other cells can show modification in their behavior. Yet we need not hesitate to accept that the cleavage cells have already diversified in their morphogenetic program. Unfortunately nothing is known about the subtle differences in the properties of cells that undergo the morphogenetic movements during gastrulation and subsequent events of development. Future research will certainly reveal the nature of "differentiation" undergone by the cleavage cells. It may not be hazardous to guess in the meanwhile that the differentiation consists of changes in the locomotory apparatus of the cells regulated by the surface membrane molecules. It is also conceivable that the differences are in the ability of the cells to respond to signals emanating from the neighboring cells and their extracellular products.

Developmental biologists have favored the study of certain simple organisms wherein "ontogenic" changes are far simpler than in the embryos of typical animals. In general, it appears that certain fundamental mechanisms of the vital activities of organisms have been conserved throughout evolution, and this has strengthened the hope of obtaining knowledge from the study of the simple organisms that lend themselves to easy experimentation. In this chapter, we shall describe some studies on the primitive eukaryotes called cellular slime molds.

A. Taxonomic Status and Affinities of the Cellular Slime Molds

Some primitive forms of life offer considerable difficulties in determining their affinities with other organisms and hence in adopting a suitable scheme of their classification. The cellular slime molds and some other allied organisms have a vegetative phase of the life cycle consisting of unicellular individuals that aggregate into larger structures in the next phase. Early biologists included them under the group Mycetozoa, signifying thereby affinities with fungi and animals. The synonym, Myxomycetes, indicates affinities with plants. Whether they are allied to plants or animals has been a continuously debated question because of the presence of features indicating affinities with both. Free amoebae of the cellular slime molds are holozoic (i.e., they ingest solid food) and have a plasma membrane similar to that found in animal cells. They also store glycogen. In fact, more than 90% of the total carbohydrates of the amoebae is glycogen. These features are adequate "qualifications" for inclusion in the Animal Kingdom.

On the other hand, during the multicellular phase of their life, the stalk and spore cells have cellulose, which is a characteristic feature of plants. Whittaker[1] has suggested that the problems arising out of having to assign all organisms to either of the two kingdoms (plants and animals) can be overcome by a new scheme consisting of five kingdoms: (1) Monera (prokaryotes), (2) Protista (chiefly Protozoa), (3) Fungi, (4) Plantae (multicellular plants), and (5) Animalia (multicellular animals). According to this scheme, the Myoxomycetes are placed under Protista. The subject can, however, be debated endlessly, and it is expedient to leave it alone in a book not directly concerned with examining the affinities of these organisms.

The cellular slime molds are undoubtedly eukaryotes with a distinctive nucleus. The messenger RNA, when first transcribed, is larger than the processed messenger and is provided with a poly-A tail added posttranscriptionally at the 3' end.[2] However, their genome size is one of the smallest among the eukaryotes, just about five times the size of *Escherichia coli* genome. Older schemes of classification included two distinct types of slime molds under the same group, viz., Mycetozoa or Myxomycetes: (1) forms that are unicellular and amoeboid in the vegetative phase, becoming multinucleate syncytia later in the life cycle, and (2) the forms that have a vegetative phase of unicellular amoeboid individuals which aggregate and form multicellular structures later. Since cell fusion and formation of a syncytial phase never occur in the latter category, they have been considered a distinct group, Acrasiales, or the cellular slime molds, distinguished thereby from the true or syncytial slime molds. The cellular slime molds are also characterized by the absence of flagellate cells and any form of sex cells during their life cycle. The cellular slime molds, which have been studied extensively, belong to the family Dictyostelidae that includes the following genera: *Dictyostelium, Polysphondyleum, Cenonia*, and *Acytostelium*. For a detailed discussion on the classification of these and allied organisms, reference may be made to Olive.[3] For early literature on the occurrence, collection, culture, and other aspects of the cellular slime molds, the works of Olive[3,4] may be consulted. The developmental aspects of *D. discoideum* are summarized in a book edited by Loomis.[5]

II. THE LIFE CYCLE OF CELLULAR SLIME MOLDS

Before we describe the experimental work appropriate to the main theme of this book, it is necessary to describe briefly the morphology and life history of the organisms, particularly since they are not usually included as "types" in introductory courses of zoology or botany.

Dictyostelium discoideum has been by far the most extensively studied cellular slime mold. The vegetative phase consists of unicellular amoebae, which occur in moist soil in humid places. They have lobopod type of pseudopodia. They ingest a variety of bacteria and multiply rapidly. When the food organisms are exhausted, the amoebae show a peculiar behavior of migrating towards a center where they aggregate into a multicellular mass that is called a pseudoplasmodium (to distinguish it from the plasmodia of true slime molds) and subsequently becomes an upright grex. Aggregation of the starved amoebae results from their chemotactic response towards the aggregation center. A slimy mucopolysaccharide covering is secreted around the grex. The total number of cells in a grex depends largely on the number of amoebae that come under the influence of the aggregation center. Larger pseudoplasmodia developed from a dense colony may contain many thousand cells, whereas small ones may have just a few hundred.

In response to certain external stimulations, the grex can show a remarkable locomotory activity. The grex showing the characteristic migratory activity is therefore called a "slug". During the migration, the slimy sheath around the cellular mass remains stationary and is left behind as a trailing collapsed tube. Grex migration can continue even up to 20 days. The migratory activity of the slug is influenced by the environmental conditions. Low pH and ionic strength encourage migration. Gradients of temperature, humidity, and light orientate the migrating slug. When light falls on the surface from a side, the slug moves towards

FIGURE 1. Life history of the cellular slime mold *Dictyostelium discoideum*. The complete life cycle takes about 24 hr under certain conditions in the laboratory. The approximate time (hours) taken for transformation from one stage to another is indicated by encircled numbers beside thick arrows. The thin arrows show the sequence of stages. The loop at the top right illustrates the germination of spores and multiplication of the amoebae. The loop at the bottom left illustrates the migration of the slug, which may be prolonged, abridged, or even dispensed with. The stages are not drawn to scale.

it. Light from the top stops migration. Obviously these features are adaptations related to the distribution and dispersal of the species.

After the migratory activity (or even without it), the grex eventually undergoes morphological changes collectively called "culmination" and gives rise to two distinct parts, the stalk and the spore-bearing tissue. From the anterior, approximately one third part, the stalk is differentiated while the remaining tissue forms the spore, producing a structure called a fruiting body. The fruiting body dries and can survive prolonged periods of drought. When the dispersed spores find themselves in a favorable environment, they germinate, and vegetative amoebae are liberated from them. The life history of *D. discoideum* is depicted diagrammatically in Figure 1. The approximate time taken by the various stages of life cycle under standardized culture conditions is also indicated in the diagram. Migration of the slug is not an essential step in the life cycle. Its duration varies considerably, depending on the external environment.

Different species of the cellular slime molds show variation from *D. discoideum* in some of the features described above. Aggregation of *D. discoideum* amoebae occurs through their migration in characteristic streams, whereas in *D. polycephalum*, it is by the migration of cells as sheets. The migration of grex is observed in *D. polycephalum* but not in *D. mucoroides*, *D. lacteum*, and *D. minutum*. The morphology of the fruiting bodies is also characteristic and often diagnostic of the species (Figure 2). Some possess characteristic colors (e.g., purple: *D. purpureum*, *P. violaceum*; white: *P. pallidum*; yellow: *D. discoideum*, *D. mucoriodes*, etc.). In *D. lacteum*, a number of mounds are formed from the aggregated cells, and eventually a fruiting body is formed from each mound. In *D. polycephalum*, several stalks are held together forming a composite fruiting body. *Polysphondyleum* fruiting bodies consist of a main stem with whorls of stalks bearing spores. These differences are mentioned to suggest that morphogenesis in the cell aggregates is highly organized, and there must be definite mechanisms which control the detailed patterns of development. The nature of these mechanisms, however, remains entirely unknown. Ex-

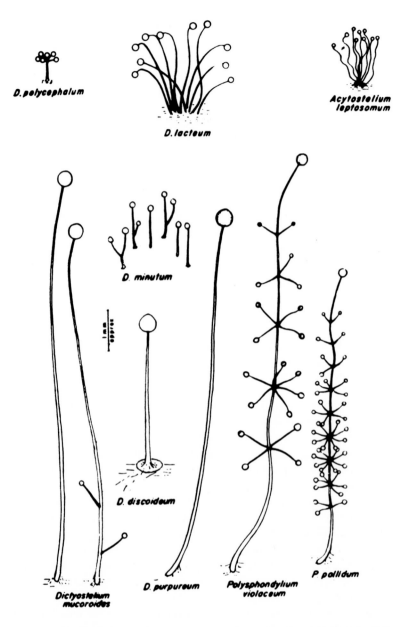

FIGURE 2.   Semidiagrammatic drawings of the fruiting bodies of some species of cellular slime molds. (From Cavender, J. C. and Raper, K. B., *Am. J. Bot.*, 52, 302, 1965. With permission.)

perimental studies on species other than those already investigated will be highly rewarding in this context.

Experimental studies using the cellular slime molds have been facilitated because they can be grown in culture. Bacteria are provided as food organisms for the growing amoebae that are grown on agar coated petri dishes. A wide variety of bacterial species are acceptable to the amoebae, though it has been observed that different species of the slime molds show some food preferences. A single species of bacteria grown in culture can be offered as food to cultures of the amoebae. *P. pallidum* and *D. discoideum* have also been grown as axenic cultures. In general, axenic cultures grow slowly and differ in some other respects from those grown on bacterial food.[6]

In the life history of the cellular slime molds, the change from the vegetative to a social phase is of considerable significance. It offers an opportunity to investigate cell adhesion and migration in a comparatively simple system. During the vegetative phase, the amoebae divide repeatedly, but do not show any tendency to adhere to each other. They show typical amoeboid movement without exhibiting any collective behavioral pattern. After the food is withdrawn, their behavior changes. Discrete centers of attraction are established by specialized cells, which attract other cells around them. The responding amoebae orientate towards the center and begin to move in streams. The streams join with others like the tributaries of a river, and the larger streams thus formed migrate towards the center where the cells pile up into several layers and ultimately an upright grex is formed. The aggregation-competent amoebae differ from the vegetative ones in at least two respects: the former are strongly adhesive and responsive to the chemotactic stimulus emanating from the reaggregation center, whereas the latter are not. The change from a nonadhesive to a highly adhesive state offers the opportunity to investigate cell adhesion in general. Similarly, the chemotactic response and the definite pattern of migration exhibited by the aggregating amoebae constitute a simple system of collective cell behavior, akin to the organized morphogenetic movements of embryonic cells.

III. CELL ADHESION AND RECOGNITION

The vegetative amoebae are not entirely nonadhesive cells. In stirred suspensions, they form cell clusters. This resembles the mutual adhesion of previously disaggregated cells of vertebrate embryonic organs in agitated suspensions. Aggregation of the vegetative amoebae is completely inhibited by EDTA. In contrast, the aggregation-competent (i.e., starved) amoebae show an EDTA-insensitive adhesion, which probably depends on a mechanism distinct from that possessed by the vegetative cells. Here we have a situation of Ca^{2+}-dependent and independent mechanisms reminiscent of those discussed in Chapter 2 of this volume. Adhesion of cells obtained from mechanically disaggregated grex is Ca^{2+}-independent, but seems to depend on intact proteins of the surface since protease treatment of the amoebae renders them nonadhesive.

Attempts have been made to discover the qualitative and quantitative changes in surface proteins related to the transition from the vegetative to the aggregating phase. A complex change involving many proteins has been described: some new proteins appear and some diminish or disappear. Das and Henderson[7] have reported that there are highly stage-specific proteins that are developmentally regulated. However, there is nothing to indicate which of them constitutes the adhesive mechanism. The work of Gerisch et al.[8] indicated that there may be two types of cellular contact sites, A and B. The contact sites A are characteristic of the aggregation-competent cells, whereas the sites B are present in the vegetative as well as aggregation-competent amoebae. Monovalent Fab fragments of antibodies against the contact sites revealed another interesting aspect of the adhesive mechanism. Blocking contact sites B with the respective Fab prevented side to side adhesion. Aggregating amoebae adhere side to side and also end to end (Figure 3), whereas the vegetative amoebae adhere only side to side. Thus the two mechanisms are apparently related to the two ontogenically distinct adhesive patterns. Garrod et al.[9] discuss some of their experimental results, indicating that the adhesiveness of vegetatively growing *D. discoideum* amoebae is diminished by an inhibitory factor secreted by them. The inhibitory factor is a low molecular weight substance (between 500 and 700 mol wt) and presumably binds the adhesion sites B. A consequence of this would be the prevention of clumping of the amoebae in their vegetative phase of the life cycle. After aggregation, the inhibitor has no effect on development. The inhibitor is not species specific: it is effective against the aggregation of vegetative cells of three other speices, *D. purpureum*, *D. mucoroides*, and *P. violaceum*.[9] The adhesion mechanism of *D. discoideum* is more complex and includes additional components.[10,11]

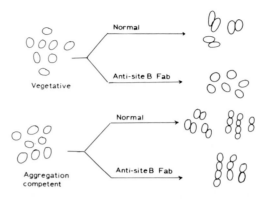

FIGURE 3.   The effect of antisite B Fab on the reaggregation of vegetative and aggregation-competent amoebae. See the text.

An important aspect of cell adhesion in the cellular slime molds is the existence of a species-specific adhesive mechanism. Amoebae of different species may be living in the same soil in nature though they do not form mixed aggregates and fruiting bodies. This is not solely due to the existence of different chemotactic agents in the various species. Even in the species of *Dictyostelium* that respond to the same chemotactic agent, separate fruiting bodies are formed from a mixed population of amoebae though an initial mixing may take place.[12] It is known that cell recognition is effected through cell surface molecules. In fact, the cellular slime molds have offered a very convenient biological material to study the molecules involved in cell recognition. Cell adhesion and detachment can be studied using columns of hydrophobic beads.[13] Suitable modifications of the technique can be employed for experimental evaluation of the different known adhesive mechanisms.

A. Studies Using Plant Lectins

Plant lectins have been used as membrane probes in studies on the adhesion mechanisms of the amoebae. Con A and WGA agglutinate vegetative phase amoebae more effectively than the starving ones.[14] However, the number of Con A receptors does not diminish with the developmental change.[15,16] Thus the decreased adhesiveness is not likely to be due to a decrease in the number of contact sites, but rather due to qualitative changes in them involving alterations in the composition of the oligosaccharide chains of the surface glycoproteins. Membrane fluidity can also be implicated in the changing adhesiveness of the cells. Von Dreele and Williams[17] have measured membrane fluidity at various stages of the development of *D. discoideum*. No change was found in this property, and this finding supports the tentative suggestion about the possible alteration in the oligosaccharides of the Con A receptor. Definitive evidence on this is not yet available. Ivatt et al.[18] have recently investigated the pattern of glycoprotein synthesis during the development of *D. discoideum*. However, more information is needed to offer a comprehensive account. From an earlier chapter, we may recall that cell surface glycosyl transferases can add monosaccharide residues to already existing oligosaccharide chains of plasma membrane glycoproteins or glycolipids. Decreased agglutinability of the starving amoebae in the presence of Con A suggests that cell recognition properties can alter through the changes caused by alterations in surface carbohydrates. Diminished agglutinability may also be due to the association of lectin binding sites with restricting cytoskeletal structures.

Effects of plant lectins have also been studied on the development of *D. discoideum*.[16] It was observed that Con A treatment delays the onset of aggregation. Interpretation of this observation is, however, difficult since the effect of the lectin could be complex. It cannot be concluded that the delay in development is due to binding of Con A to the cell adhesion mechanisms. Quite like lymphocytes, the *D. discoideum* amoebae show lateral shifting of membrane bound Con A, and "capping" of the ligands occurs followed eventually by endocytosis. The effect of the lectin thus includes a wide spectrum of membrane alterations, and the delay in development cannot be explained easily. Possibly the delay corresponds to the time taken for the resynthesis of some essential membrane components, which are internalized or otherwise altered following the lectin binding. The studies using plant lectins have, however, established the fact that complex cell surface changes are the basis of transition from the vegetative to the aggregating phase of the amoebae.

A peculiar relationship between the WGA receptors of *D. discoideum* and the effect of yeast phagocytosis has been discovered by Hellio and Ryter.[19] After the amoebae have phagocytosed yeast, the WGA receptors on the surface disappear. This does not happen when bacteria or intact latex beads are phagocytosed or when droplets of axenic media are pinocytosed by the amoebae. This shows that there is a specific relationship between yeast cells and WGA receptors. Whether the WGA receptors have any adaptive value to the amoebae is not clear. It is, however, quite likely that the amoebae have surface mechanisms that recognize the food organisms by a lectin-receptor system and thereby exercise their choice.

B. Endogenous Lectins of the Cellular Slime Molds

A novel avenue of approach to understanding the development of cellular slime molds has been opened by the studies of Rosen and Barondes and associates.[20,21] They have demonstrated that these organisms have their own lectins, which play a significant role as recognition molecules. Extracts of the amoebae at different stages of development (time after withdrawal of food) were tested for the hemagglutinin activity. Sheep erythrocytes showed agglutination in the presence of some of the extracts, thereby demonstrating that the cells possess developmentally regulated sugar binding proteins. The assay consists of taking the extract in serial dilution (twofold or tenfold dilution at each step) in the depressions of suitable test plates and adding a standard volume of a suspension of rabbit or sheep erythrocytes. The erythrocytes may be previously treated with formaldehyde or glutaraldehyde. After incubating for an hour or so, the erythrocytes agglutinate if the concentration of the agglutinin is above a certain level. Higher dilutions do not show agglutination. The assay gives a semiquantitative estimate of the lectin in the extract. Using the same assay procedure, one can determine the sugar to which the lectin has binding specificity. Again, a set of depressions is filled with the sugar to be tested in serial dilutions. The erythrocytes and a constant amount of the lectin (which is just enough to bring about agglutination) are also present in the test depressions. Sugars, for which the lectin has binding specificity, inhibit competitively the agglutination of erythrocytes at considerably low concentrations; those without specificity cannot inhibit at all, or do so only at very high concentrations. Rosen et al.[22] obtained extracts from aggregation-competent amoebae of six species and tested their ability to agglutinate erythrocytes. Competitive inhibition of hemagglutination indicated that the extracts contain lectins with specific sugar binding characteristics. It is known that lectin binding specificity towards more complex sugars is greater than towards mono- or disaccharides. So we cannot conclude that the lectin binding site has been identified; however, the sugar that inhibits hemagglutination is most likely to be the essential part of the molecular domain where binding takes place.

Purification and characterization of the lectins has been achieved in the case of *D. discoideum* and *P. pallidum.*[20] When the extract of *D. discoideum* amoebae was passed through a column of Sepharose® 4B, the lectin was retained. Sepharose® is a linear polymer of galactose, and the retention of the lectin is due to its specific binding. Elution by free galactose yields purified lectins. By a further step of affinity chromatography on DEAE cellulose, the lectins could be resolved into two fractions called discoidin-I and discoidin-

II. The sugar binding specificity of the two discoidins is the same, but they differ in other respects. The two discoidins seem to be distinct proteins, i.e., products of different genes as indicated by differences in their tryptic peptides. Slight differences between the two are also indicated by the relative ability of simple sugars to inhibit erythrocyte agglutination. They also differ in their agglutinating activity with respect to rabbit, sheep, and human erythrocytes. Further they differ in the time of their appearance and rate of increase during the differentiation of the vegetative amoebae into aggregation-competent ones.[23] Discoidin-I level begins to rise sharply at about 6 hr from the time of food withdrawal and reaches a 50-fold level in the next 6 hr. Discoidin-II also begins to rise at about the same time as discoidin-I, but does not reach a very high level. Whether these differences have a regulatory role is, however, not clear.

Similar investigations on *P. pallidum* have yielded information that lends itself to comparison with the discoidins. The lectin from *P. pallidum* has been named pallidin. It is distinct from the discoidins: its pI is close to 7.0, whereas that of the discoidins is about 6.1.[24] Pallidin has been purified by adsorption to formalinized human group O erythrocytes. It is eluted from the erythrocytes by the addition of free galactose.[24]

For an effective role in selective cell adhesion, the lectins have to act at the cell surface. Several observations have established that the cellular slime mold lectins act at the cell surface. They can be extracted easily from intact amoebae using media containing the appropriate sugars, which bind them specifically. Chang et al.[25] demonstrated that fluorescent or ferritin conjugated antibodies against rabbit immunoglobulin could label cells coated with rabbit antidiscoidin; no binding was observed in vegetative amoebae. Similar observations have been made on pallidin.[26] Labeling is inhibited by the presence of competitive sugars. It is also observed that aggregation-competent amoebae form rosettes with formalinized sheep erythrocytes, whereas the vegetative amoebae cannot do so.[26] Recent work of Barondes and group[27] has clarified several points of uncertainty in the interpretation of experimental results, implicating a role for lectins in the morphogenesis of cellular slime molds. The discoidin binding sites at the cell surface have recently been localized.[28] A distinction has been made between membrane-located lectins and the so-called soluble lectins. The latter are secreted by the cells into the intercellular space. They are obtained without the use of detergents to solubilize the plasma membrane. The discoidins and other slime mold lectins are of the soluble type.[27]

During early aggregation of the amoebae, discoidin-I is mostly intracellular. In the formed aggregates, it is externalized by the cells. It is associated with the slime coat of the mature aggregates. This suggests that discoidin-I does not act as a cell-cell ligand. This conclusion is further fortified by the finding that antibodies to discoidin-I do not block the adhesion of the amoebae in an in vitro assay.[10] Discoidin-II is also initially intracellular. Subsequently, it is externalized around the differentiating spore cells.[27]

Synthesis of lectins by the cellular slime mold amoebae seems to be directly related to the attainment of aggregation competence and triggered by withdrawal of food. An apparently anomalous situation is created by the finding that axenically cultured vegetative amoebae of *D. discoideum* show traces of discoidins and possess the EDTA-insensitive aggregation mechanism. However this may be due to the peculiar condition of culture. Perhaps the amoebae are "half-starved" in the unwholesome medium and hence show some characteristics of the aggregation-competent amoebae. In case of amoebae feeding on bacteria, starvation triggers the synthesis of the lectin. Withdrawal of food is followed by the synthesis of fresh mRNA for the synthesis of lectins within a couple of hours. Actinomycin treatment inhibits the synthesis of lectins. Glucose starvation does not seem to have any effect on the initiation of the synthetic activity. Starvation of amino acids seems to be the critical stimulus for the appearance of aggregation competence. A relationship between mitotic activity and a high rate of amino acid uptake has been discussed by Bhargava.[29] It is likely that abundant

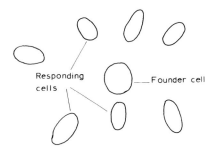

FIGURE 4.   Diagram showing the initiation of aggregation in cellular slime molds. In some species, a cell becomes rounded and becomes the founder cell. In others, the founder cell is not morphologically distinguished. In any case, the surrounding amoebae orientate to approach the founder cell.

food supply engages the amoebae in continuous mitotic cycles suppressing differentiation. A critical cell density also seems to be essential for some early synthetic activity in the differentiating amoebae, including an increase in actin.[30]

IV. AGGREGATION AND FORMATION OF THE SLUG

After the food organisms are exhausted, the amoebae begin their social phase of the life cycle. In other words, they show a collective behavior. An important aspect of this is acquiring the ability to respond to the chemotactic stimulus attracting them towards a center. Under experimental conditions, the *D. discoideum* amoebae begin to respond after about 8 hr from food withdrawal. Early workers recognized the fact that the attracting center must be secreting some chemical, which was named acrasin. It is now clear that the chemoattractant of all species of cellular slime molds is not identical. It has been established that cyclic 3′:5′-adenosine monophosphate (cAMP) acts as the chemoattractant for the amoebae of *D. discoideum*, *D. mucoroides*, *D. rosarium*, and *D. purpureum*. Amoebae of *D. lacteum*, *D. minutum*, *P. violaceum*, and *P. pallidum* are not attracted by cAMP.[31,32] Folic acid is also known to act as a chemoattractant for the cells of *D. discoideum*.[32] The acrasin of *P. violaceum* has been shown to be a heat-stable peptide of low molecular weight (<1500). This substance can attract the amoebae of *P. pallidum* also.

The initial source of the acrasin seems to be a specialized cell. It may be morphologically indistinguishable from the other amoebae as in *D. discoideum* and several other species. In some species of the cellular slime molds such as *P. violaceum*[33] and *D. minutum*,[34] it was observed that the aggregating amoebae migrate towards a specialized, morphologically distinguishable cell that acts as the founder cell. It is a highly adhesive rounded cell (Figure 4). It seems that a founder cell can arise from any of the amoebae in response to the stimulus for aggregation. This is concluded from the observation that cloning does not diminish the potency of center formation.[34] The amoeba destined to be the founder cell gradually loses its motility, withdraws the pseudopodia, and assumes a more or less spherical shape. Eventually the other amoebae around begin to orientate and move towards it (Figure 4). The founder cell can, however, turn into an ordinary cell, and thus its differentiation is reversible.[33,34] In other species too, where morphologically distinguished founder cells are not observed, it is known that a single cell constitutes the initial center of attraction. What causes a cell to differentiate into such a functionally specialized cell is not known clearly. Sinha and Ashworth[35] have described a parasexual cycle in *D. discoideum*. The vegetative amoebae and the cells in the differentiated grex and fruiting bodies are haploid (n = 7). There is no fusion of gametes or reduction division during the life cycle. However, two vegetative amoebae fuse occasionally and a diploid cell is formed. The diploid state is only transient and the "extra" chromosomes are eliminated one by one. As a consequence, the cell passes through a series of aneuploidies. It has been suggested that cells in a stage of aneuploidy with eight chromosomes are those that become the founder cells.[6]

The sequence of events in the aggregation of the amoebae of *D. discoideum* is beginning to be understood, at least in its salient features. The entire process is operated through a system of cellular communication mediated by simple chemical substances and physiological alterations in the cells. In *D. discoideum,* there is a rhythmic activity generated in the population of the amoebae responding to the chemotactic signal. When cAMP is released from the cell acting as the founder cell, it reaches the surrounding amoebae by diffusion. The responding amoebae have cAMP receptors on the surface of their plasma membrane. On binding of the chemical signal, they undergo changes leading to (1) their movement towards the founder cell, and (2) release of cAMP from their cytoplasm into the medium. The release of cAMP occurs in pulses every 5 to 10 min. Movement of the starving amoebae towards the center continues for about 100 sec during which the distance covered is about 20 μm. A second pulse of cAMP from the center attracts the amoebae further, and at the same time, the responding amoebae themselves have produced a pulse of cAMP attracting other amoebae located further away from the center. The pattern of rhythmic movement is generally of concentric rings of amoebae responding to the stimulus. However, the pattern may change to streams at low density. Concentric rings or spirals are found in high-density cultures. The response of the cells to cAMP includes marked changes in the cytoplasmic contents which are expressed as fluctuations of the optical density of bubbled suspensions of the amoebae. The period of rhythmic movement is not constant: at the start it is about 10 min, decreasing rapidly to 5 min and then gradually to 2.5 min. When the pulse of cAMP is released from the center, there is a quick response of the amoebae as described above. However the extracellular cAMP level is soon lowered as it is broken down by phosphodiesterase enzymes produced by the starving cells and presumably located on the cell surface. The enzyme is distinct from the cAMP receptor.

It has been shown conclusively that the cAMP is bound to specific receptors on the cell surface. This is known from the kinetics of binding of radioactively labeled cAMP on the cells. Whereas the responding cells of *D. purpureum, D. discoideum,* and *D. mucoroides* show definite increase in the amount of bound cAMP with its increasing concentration, the cells of *D. minutum, P. violaceum,* and *P. pallidum* do not show an increase. This difference accords with the fact that cAMP is not the chemoattractant in the case of the species that do not show the specific binding.[36] The effect of phosphodiesterase is overcome in such experiments by using excess cGMP (cyclic guanosine monophosphate) to keep the enzyme activity diverted, or by other methods.[37] Thus the observed binding of cAMP or the lack of it is not an artifact.

The behavioral response of the amoebae to the pulse of cAMP is complex and is not understood in detail. How the bound cAMP stimulates the cellular locomotory activity and determines its vectorial nature is not clear. Galvin et al.[38] have presented evidence that *D. discoideum* cells can be extracted with the nonionic detergent Triton X 100, retaining the cytoskeleton and the cAMP binding sites. Since chemotaxis should involve a concerted action of the membrane receptors and the cytoskeleton, it is quite probable that the cAMP receptors impose a polarity on the cell and thereby determine the direction of locomotion. Recent studies of Yumura et al.[39] have elucidated the role of cytoskeletal proteins in the locomotion of the amoebae. The localization of myosin in the cell cortical region may be correlated with the receptor for the cAMP.

An increased influx of extracellular Ca^{2+} observed as a consequence of the cAMP pulse may be of significance in this context. In the cytoplasm of the cells responding to cAMP, there is an increase in the level of cGMP, which shows a quick and sharp decrease within a minute. This is followed by a second peak of cGMP as well as cAMP, the latter being released into the extracellular space. It has been suggested that the release of the cAMP occurs through exocytosis, but the evidence for this is not definitive. The other biochemical oscillatory changes related to these are alterations in the redox state of cytochrome b and fluctuation in the pH of the external medium caused by the release of hydrogen ions.

The cycle of changes described above generates the characteristic rhythmic oscillations consisting of waves of excitation passing from the center through a relay system in which peripherally located cells act as amplifiers of the signal generated at the center. One of the key events in the cyclic changes is the release of cAMP in pulses. The sequence of events following the binding of cAMP on the receptor can be explained on the assumption that the receptor acts in association with guanylate cyclase, which increases the intracellular cGMP level. The elevated cGMP activates adenylate cyclase, resulting in a peak of intracellular cAMP, which is released out. A comprehensive model that takes into account most of the experimental observations has been presented by Gerisch et al.[37]

V. CELL INTERACTIONS IN THE DIFFERENTIATION OF THE GREX

The cells constituting the grex differentiate into either of the two broad types — stalk or spore cells. In the posterior two thirds portion of the grex, the cells differentiate into the spores and in the anterior one third into the stalk. This does not seem to depend on the total number of cells in the grex; whether it is a large or small cellular mass, the ratio of the two types of differentiated cells remains the same. A more interesting observation is that the grex may be operated on to remove a part. Again the anterior third of the remaining cellular mass develops into the stalk cells and the rest into spores. Evidently the ''determination'' of the cells is reversible at this stage. It cannot, however, be concluded that all the amoebae constituting the grex are identical.

Leach et al.[40] have demonstrated that cells grown in media containing glucose and media lacking it can sort out when mixed. They differentiate into distinct types: those grown in the presence of glucose become spore cells, the rest becoming the stalk. What is the nature of the intrinsic differences in the cells which sort out? It has been suggested that the difference in their adhesiveness leads to sorting out in a manner analogous to what is described in the cells of vertebrate embryonic tissues.[41] This is, however, not a wholly satisfactory explanation since the pattern in the grex is not one of forming enclosed and enclosing phases. Subtle differences in the state of differentiation may exist in the cells that have been shown to sort out. Sorting out has been demonstrated in mixed aggregates of mutant and wild type cells or in the aggregates consisting of cells grown under different conditions of culture. Shuffling and sorting out of the grex cells has been demonstrated by using vitally stained cells also.[42] If vitally stained posterior cells of a slug are grafted onto the anterior part of an unstained slug, they lag behind during movement and eventually assume a posterior position. On the other hand, if vitally stained anterior cells are grafted posteriorly on an unstained slug, they shift to an anterior position. The nature of differences between the prespore and prestalk cells is, however, not yet understood clearly. Nevertheless, it is likely that important differences are associated with the cell surface.

Two distinct mechanisms bringing about regional differentiation can be visualized. These are depicted diagrammatically in Figure 5. It is not easy to settle in favor of one or the other with the information available at present. Garrod et al.[9] have reported experimental evidence to show that some sorting out of already differentiated cells occurs in the aggregate before the formation of the grex with its characteristic tip. Amoebae of *D. discoideum* aggregated in stirred suspensions show a random arrangement of prespore cells. These can be detected in the sections of such aggregates by fluorescent antibodies raised against the spores of another species, *D. mucoroides*. On keeping the aggregates for some time, the cells seem to sort out.[43,44] On this and other evidence, Garrod et al.[9] suggest the following.

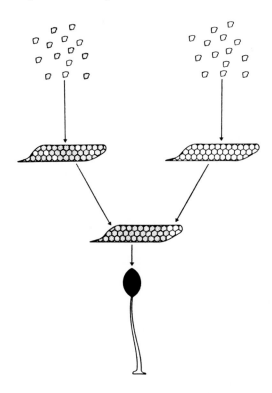

FIGURE 5.   Diagram showing two alternative ways in which a slug of *Dictyostelium* could differentiate. The left hand side represents the view that distinct cell types, which may be predisposed even before aggregation, are present in the grex, randomly distributed initially. Subsequently they sort out, forming the prestalk and prespore regions. The right hand side illustrates the view that the cells of the grex differentiate *in situ* and thus give rise to the prespore and prestalk regions. (From Garrod, D. R., Swan, A. P., Nicol, A., and Forman, D., Cellular recognition in slime mould development, *Symp. Soc. Exp. Biol.*, , 32, 173, 1978. With permission of the publisher, the Company of Biologists, Cambridge University Press.)

1.   The two cell types differentiate at random within the early grex and then sort out to give the prespore-prestalk pattern.
2.   Differentiation, sorting out, and pattern formation begin before the formation of the grex tip. Thus the grex tip does not play an organizing role in any of these processes.
3.   The early prestalk region itself gives rise to the grex tip. It is the formation of the pattern that generates the polarity of the grex tip rather than the polarity that gives rise to the pattern.
4.   Control of the ratio of prestalk and prespore cells may be independent of the mechanism that determines the spatial arrangement of cell types.

The observation that the ratio of prestalk and prespore cells is reestablished after experimental alteration necessitates the assumption that loss of differentiation and redifferentiation can occur in the grex cells. It is not surprising to find such a situation. The well-known phenomenon of regeneration of the eye lens in certain amphibia may be recalled in this context. Cells derived from a highly differentiated tissue such as the iris or cornea can differentiate into a lens under appropriate conditions. Though such a phenomenon is rare in higher animals, finding it in a primitive organism need not be considered surprising. Another interesting observation may be mentioned in this context. Grex cells liberated from the slime sheath and exposed to bacteria begin to feed and revert to the vegetative phase. This remarkable flexibility in the developmental program of the cellular slime molds is undoubt-

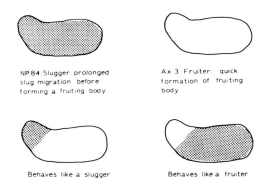

NP 84 Slugger prolonged slug migration before forming a fruiting body

Ax 3 Fruiter quick formation of fruiting body

Behaves like a slugger

Behaves like a fruiter

FIGURE 6. Altered behavior of grafted *Dictyostelium* slugs. Based on the experiments of Smith and Williams.[45] See the text.

edly of survival value to these primitive organisms. Besides, it offers a simple and versatile model for a study of the reversible state of determination.

The tip of the grex is not totally devoid of a role in development. Its supposed role in inducing the pattern of differentiation has been questioned on the basis of the evidence discussed above. However, the tip also responds to light and other external stimuli by orienting and moving in a particular direction, depending on the nature of the stimulus. Smith and Williams[45] have reported a study of two mutant strains of *D. discoideum* exhibiting two distinct properties: one of the strains (NP 84) shows prolonged slug migration and hence is called a *slugger,* whereas the other (Ax3) develops into the fruiting body after a short period of migration and hence is called a *fruiter.* In transplantation experiments (Figure 6), it was shown that the switch from migration to fruiting body differentiation resides in the tip. When the tip of a *slugger* is transplanted on the posterior part of a *fruiter,* the composite slug performs prolonged migration before differentiating into the fruiting body. On the other hand, the *fruiter* tip grafted on the posterior part of a *slugger* causes the development of the chimera to proceed to fruiting body formation without exhibiting migration. On the basis of these findings, we may conclude that the tip of the grex also has an important role in development. Whether this is restricted to controlling the movement of the grex or to any additional process is not easy to conclude. One hopes that further work will clarify the situation. It must be emphasized that understanding the control of differentiation in the grex is important since it is an attractive, simple model of morphogenesis. Information regarding this will hopefully contribute to understanding more complex developmental processes.

When a grex is disaggregated and allowed to reaggregate under suitable conditions, the cells have the ability to undergo morphogenesis quickly, tending to catch up with the time schedule of development and passing through the normal morphological stages of development (see Figure 7). The recapitulation of developmental program is more rapid when the grex is disaggregated later in its development. For example, a grex disaggregated at 12 hr stage of development proceeds quickly through the morphological stages and "makes it" to the stage of fruiting body almost by the same time as an undisturbed control. If the grex is disaggregated 15 or 18 hr after starting development, the cells reaggregate and reach the ultimate stage just 2 to 4 hr late.[46] It would appear that the amoebae have already undergone considerable chemodifferentiation at the time of disaggregation and therefore can follow the normal course of morphogenesis rapidly. A grex finishing all the morphogenetic events of development in time, in spite of having to start all over again 12 hr behind schedule (which is half the total time required for development), is certainly no small performance. What are the events of development completed in the initial 12 hr? What precisely is lost after disaggregation? How are events of morphogenesis completed in the normal sequence (albeit

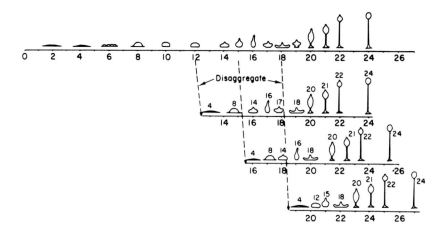

FIGURE 7.   Catching up with development by the slime mold amoebae disaggregated from a developing grex. The upper figures represent normal developmental stages in the undisturbed controls. Below these are the morphological stages observed during rapid recapitulation of development following mechanical disaggregation of the grex after 12, 15, and 18 hr of normal development. The large numbers below the figures represent the time at which that particular morphological form is reached in the undisturbed controls. (From Newell, P. C., *Essays Biochem.*, 7, 87, 1971. With permission.)

considerably faster) in spite of starting as cells that had once gone through a part of the developmental program? These questions remain to be answered.

Aggregation-competent amoebae synthesize certain enzymes, which are developmentally regulated. In *D. discoideum* amoebae, the enzyme UDPG pyrophosphorylase (EC 2.7.7.9) begins to be accumulated in the spores and stalk tissue at about 12 hr after withdrawal of food. The increase, as well as the subsequent decrease, in the enzyme activity is dependent on concurrent protein synthesis and prior RNA synthesis. Inhibition of RNA synthesis by actinomycin D at 15 hr or later does not prevent the accumulation of the enzyme in intact grex tissues, indicating that the messengers are considerably stable. If the grex is disaggregated at this stage, the amoebae can aggregate quickly and catch up with the morphological developmental changes as mentioned above. Besides, the UDPG pyrophosphorylase begins to be synthesized. The amount of the enzyme synthesized is the same as in normal development and regardless of the quantity present already. Up to three times the quantal increase of the enzyme has been observed on redispersion of the grex cells. The fresh synthesis of the enzyme after reaggregation is inhibited by actinomycin D, indicating that fresh RNA synthesis is essential. When the disaggregated amoebae are not allowed to reaggregate, the enzyme accumulation stops abruptly, suggesting that assuming a multicellular condition is essential for the synthesis of the enzyme.[47] Similar quantal rise has been demonstrated in the case of some other enzymes, viz., UDP-Gal-4-epimerase (EC 5.1.3.2), trehalose-6-P synthetase (EC 2.3.1.15), and UDP galactose:polysaccharide galactosyl transferase.[47] All these additional rounds of enzyme synthesis require new periods of RNA synthesis (Figure 8).

A curious property of the cellular slime molds has been revealed while studying the development of mutants. A variety of mutants of *D. discoideum* have been obtained. They show various developmental defects. A mutant of which the amoebae are unable to aggregate can be enabled to aggregate and complete development by mixing some cells of the wild type or of another mutant that can aggregate. The spores produced from such "cooperative" development consist of both the types combined, thus showing clearly that the aggregationless mutant is not defective in other traits required to complete development. As little as 10% of the total cell mass constituting the "corrective" type may suffice to lead the other amoebae

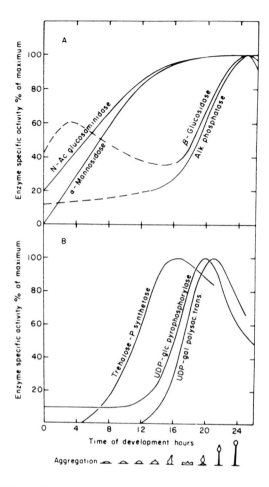

FIGURE 8.   Changes in specific activities of some of the enzymes formed during the development of *Dictyostelium discoideum*. (A) Catabolic, and (B) polysaccharide synthetic enzymes. The activities shown by broken lines are isomers of the latter enzymes. (From Newell, P. C., *Essays Biochem.*, 7, 87, 1971. With permission.)

through completion of development. The mechanism of "cooperative" development is, however, not clear. Possibly the mutant is unable to respond to the cAMP signals, but is otherwise normal.[46]

Mutants exhibiting defective mechanism of cell adhesion are of particular interest, since they lend themselves to experimental studies on the control of adhesiveness. Wilcox and Sussman[48] describe a mutant, JC-5, of *D. discoideum* that can develop normally if maintained at 22°C. At 27°C, it develops normally to a postaggregative stage, but then disperses into single cells. The cells can develop into stalk and spore types, suggesting that the mutation has not altered the genotype much. It has been suggested that some protein, presumably a component of the plasma membrane, has undergone some change in the mutant, rendering it thermolabile so that the cells fail to be held together. Even at the permissive temperature (i.e., at 22°C), the cells can be rendered nonadhesive by treatment with cycloheximide, thus indicating that the synthesis of some new protein is essential to hold the cells together though the parent strain is not similarly cycloheximide-sensitive (see Figure 9). This difference in cycloheximide sensitivity suggests that the simple explanation of a thermolabile component required for cell adhesion is not adequate. It is known that the defect is not associated with

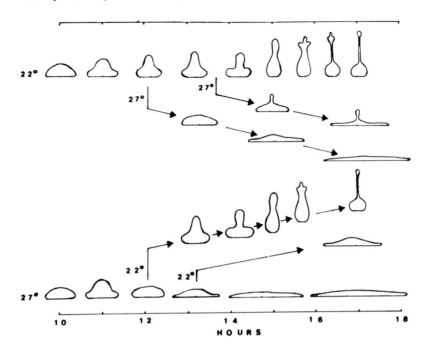

FIGURE 9.   The morphogenetic properties of the mutant JC-5. Cells were grown, harvested, washed, and dispersed on filters in aliquots of 5 × 10⁷ cells. Incubation temperatures were as noted in the diagram. At the times indicated by the abscissa, schematic drawings were made of the representative structures viewed at 20 × magnification in dissection microscope. (From Wilcox, D. K. and Sussman, M., *Dev. Biol.*, 82, 102, 1981. With permission.)

the discoidins or the contact site A. Certainly more needs to be learned about the mechanism that fails in the mutant at the higher temperature.

Several models of pattern formation in the *D. discoideum* slug have implicated cAMP as a "morphogen". For example, Rutherford et al.[49] have proposed a model consisting of a cAMP source at the prespore end, and a sink (the enzyme phosphodiesterase) at the prestalk end of the slug. This would generate a gradient of cAMP, which is assumed to regulate the cell-type differentiation. The precise manner in which the gradient of cAMP can play a dual role of spore and stalk specification is, however, not clear. It is known that cAMP controls the levels of several mRNA species that code for developmental proteins.[50] In the undisturbed slug cells, both "developmental" and "housekeeping" mRNAs are relatively stable, with a half-life of several hours. When the slug cells are disaggregated, the half-life of the developmental mRNAs decreases to about 30 min, the housekeeping mRNAs remaining unaffected. If cAMP is added to the isolated cells, the effect is reversed.[51] A complex mechanism involving controlled turnover of different species of mRNA can explain the origin of the spore-stalk dichotomy in the differentiation of the slug cells. The information available at present is, however, insufficient to elaborate this. Some additional "inducing" substance also seems to have some role in the differentiation of spores.[52] It may be added here that the slug consists of not only the two cell types (prestalk and prespore), but also an additional one, named "anterior-like",[53] dispersed throughout the slug. This can add a new dimension to the complication of any model that seeks to explain differentiation on the basis of concentration gradients of substances. In any case, it is clear that cAMP is not a mere chemotactic signal;[54-56] it dictates some events of cell differentiation also. More recent work using isolated cells[51] has confirmed the earlier suggestion that cAMP induces both stalk and spore cell differentiation. However, it is pathway-indifferent, and the choice between developmental pathways is exercised using two distinct endogenous morphogens.

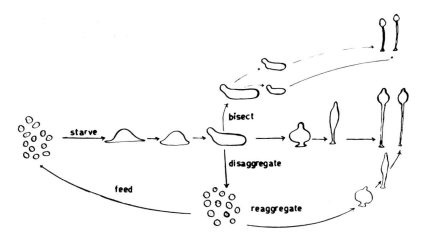

FIGURE 10. Developmental versatility of the cellular slime mold, *Dictyostelium discoideum*. The straight horizontal arrows from left to right indicate the normal course of development. The outcome of experimentally imposed alteration is indicated by the other arrows.

Stalk formation is favored by a substance called differentiation-inducing factor (DIF). It has been shown that the DIF is a low molecular weight substance (dialyzable; molecular weight <500) and is active at very low concentrations.[57] Ammonia (at millimole concentration) antagonizes the action of DIF and favors the differentiation of spores.[58]

VI. SOME SALIENT FEATURES OF THE DEVELOPMENT OF CELLULAR SLIME MOLDS

Everyone will agree that the development and life cycles of the cellular slime molds are remarkable biological phenomena in their own right, and their study is richly rewarding in itself. The literature on the cellular slime molds is now quite voluminous, and *D. discoideum* can be rightly called a famous organism. The most attractive feature of the cellular slime molds is their versatility and adaptability in development. Some aspects of this are depicted in Figure 10. Experimental manipulation is greatly facilitated by this feature. Undoubtedly, the cellular slime molds have thrown open a very interesting research field. However, we have justified including a study of these organisms in this text on another score. It is the assumption that it can reveal the mechanisms underlying the development of animal embryos. Most investigators studying the cellular slime molds have emphasized that their studies could yield valuable clues to understanding more complex developmental phenomena encountered in the higher forms of life. In other words, these simple organisms are studied as ''model'' systems to elucidate developmental processes of higher organisms. This hope is based on the broad generalization that some basic principles of life activities have undergone little change in the process of evolution. One may mention such mechanisms as cell locomotion, cytoplasmic contractility, and perhaps some others as examples. Let us therefore examine if any generalizations bearing on the development of higher organisms have emerged from the study of the cellular slime molds.

In the embryos of animals, there are numerous instances of discrete cells coming together and condensing into a more compact tissue. In the vertebrate embryos, cells arising from the sclerotome condense, laying down the rudiments of the vertebrae. The definite pattern of their migration is presumably due to some morphogenetic, perhaps chemotactic, signal arising from the notochord. It is of course equally possible that the cells move away from the somite due to some repulsive force. A combination of the two processes also cannot be ruled out. Discrete migrating cells such as the neural crest and primordial germ cells probably possess mechanisms to recognize the route as well as the destination during their complex journeys. The nature of any chemical signals that may be proposed in working hypotheses to explain these processes is not easy to discover. The complex tissue environment in which the cells migrate in the embryos of higher animals renders the identification of such hypothetical signals extremely difficult. Amoebae of the cellular slime molds offer a considerably simplified situation. Release of a simple substance by one of the cells, its binding on receptors possessed by the responding cells, and the production and release of the same substance by the latter can bring about a pattern of morphogenesis. This information, obtained from the slime molds, indicates that migration of cells towards an attracting center can work with a limited number of controlling elements. One can therefore guess that examples of discrete cells migrating in embryos could also be controlled by similar but probably more complex mechanisms.

Gastrulation movements constitute a much more complex morphogenetic process involving direction and extent of cell motility. The spatial pattern observed in the gastrulation movements seems to be orchestrated by more complex mechanisms. Aggregation of the slime mold amoebae is also a morphogenetic movement, albeit greatly simplified. The concentric rings and spiral patterns can be explained by rather simple control mechanisms involving periodicity in the release of the chemoattractant. Whether the gastrulation movements are caused by any chemoattractant emanating from a center has not been investigated seriously. The lesson learned from the cellular slime molds, however, suggests that such a possibility cannot be ruled out. In this context, it must be reiterated that the chemical nature of the acrasins of different species of the slime molds is not identical. Also the pattern exhibited by the aggregating amoebae may vary: there are species that show migration of continuous cell sheets and others that show individual amoebae moving in complex rhythmically pulsating patterns arising from cells passing through different phases of their response to the chemotactic agent. A mere change in cell density may bring about an alteration in the pattern of movement. Diffusion rates of the chemical signals, their affinity to the receptor and the time taken by the cells to show a response are some of the variables that generate complex patterns of migration. In the context of the slime molds, the following questions seem to be specially pertinent. Is the nature of the acrasin and its receptors related directly to the pattern of movement shown by the amoebae? Further, if two or more chemoattractants and the respective surface receptors are found together in the amoebae, will they exhibit a more complex pattern arising from an interaction of the component mechanisms? Answers to these questions are not forthcoming from extant knowledge. However, all these questions can be answered using the experimental techniques of cell biology available at present.

Cellular interactions leading to tissue differentiation have been widely recognized in embryogenesis. The classical primary organizer discovered by Spemann and Mangold[59] has remained a problem without solution, possibly because it is too complex a process to be analyzed on the basis of simple hypotheses postulating "evocators" and gradients of morphogenetic substances. Admittedly the study of the cellular slime molds does not at this time offer any clues to investigating the problem of embryonic induction further. Nevertheless, it seems reasonable to hope that information obtained from simple model systems can go a long way in solving such problems. The fact that a higher level of cAMP in the tip of the grex of *D. discoideum* can play a decisive role seems to be attractive. The study of mutants regulating the tendency of the slug to migrate or differentiate into a fruiting body after a brief period of migration has been mentioned earlier. An important point of interest here is the fact that the tip can control the differentiation of the posterior part in chimeric combinations so as to deviate from the pattern controlled by its genome. We need not hesitate to consider this as analogous to embryonic induction.

REFERENCES

1. **Whittaker, R. H.,** New concepts of kingdoms of organisms, *Science,* 163, 150, 1969.
2. **Lodish, H. F., Firtel, R. A., and Jacobson, A.,** Transcription and structure of the genome of the cellular slime mold *Dictyostelium discoideum, Cold Spring Harbor Symp. Quant. Biol.,* 38, 899, 1973.
3. **Olive, L. S.,** The Mycetozoa: a revised classification, *Bot. Rev.,* 36, 59, 1970.
4. **Olive, L. S.,** *The Mycetozoans,* Academic Press, New York, 1974.
5. **Loomis, W. F., Ed.,** *The Development of Dictyostelium discoideum,* Academic Press, New York, 1982.
6. **Ashworth, J. W.,** The development of the cellular slime molds, in *Biochemistry of Cell Differentiation,* Paul, J., Ed., Butterworths, London, 1974, chap. 1.
7. **Das, O. P. and Henderson, E. J.,** Developmental regulation of *Dictyostelium discoideum* plasma membrane proteins, *J. Cell Biol.,* 97, 1544, 1983.
8. **Gerisch, G., Hulser, D., Malchow, D., and Wick, U.,** Cell communication by periodic cyclic-AMP pulses, *Philos. Trans. R. Soc. London Ser. B,* 272, 181, 1975.
9. **Garrod, D. R., Swan, A. P., Nicol, A., and Forman, D.,** Cellular recognition in slime mould development, *Symp. Soc. Exp. Biol.,* 32, 173, 1978.
10. **Springer, W. R. and Barondes, S. H.,** Cell adhesion molecules: detection with univalent second antibody, *J. Cell Biol.,* 87, 703, 1980.
11. **Springer, W. R. and Barondes, S. H.,** Evidence for another cell adhesion molecule in *Dictyostelium discoideum, Proc. Natl. Acad. Sci. U.S.A.,* 79, 6561, 1982.
12. **Raper, K. B. and Thom, C.,** Interspecific mixtures in Dictyostelidae, *Am. J. Bot.,* 28, 69, 1941.
13. **Glynn, P. J. and Clarke, K. R.,** An investigation of adhesion and detachment in slime mold amoebae using columns of hydrophobic beads, *Exp. Cell Res.,* 152, 117, 1984.
14. **Reitherman, R. W., Rosen, S. D., Frasier, W. A., and Barondes, S. H.,** Cell-surface species-specific, high-affinity receptors for discoidin: developmental regulation in *Dictyostelium discoideum, Proc. Natl. Acad. Sci. U.S.A.,* 72, 3541, 1975.
15. **Weeks, G.,** Studies on the cell surface of *Dictyostelium discoideum* during differentiation. The binding of ^{125}I-concanavalin A to the cell surface, *J. Biol. Chem.,* 250, 6706, 1975.
16. **Darmon, M. and Klein, C.,** Binding of concanavalin A and its effect on the differentiation of *Dictyostelium discoideum, Biochem. J.,* 154, 743, 1976.
17. **Von Dreele, P. H. and Williams, K. L.,** Electron spin resonance studies of the membranes of the cellular slime mold *Dictyostelium discoideum, Biochim. Biophys. Acta,* 464, 378, 1977.
18. **Ivatt, R. J., Das, O. P., Henderson, E. J., and Robbins, P. W.,** Glycoprotein biosynthesis in *Dictyostelium discoideum.* Developmental regulation of the protein-linked glycans, *Cell,* 38, 561, 1984.
19. **Hellio, R. and Ryter, A.,** Relationships between anionic sites and lectin receptors in the plasma membrane of *Dictyostelium discoideum* and their role in phagocytosis, *J. Cell Sci.,* 41, 89, 1980.
20. **Rosen, S. D. and Barondes, S. H.,** Cell adhesion in the cellular slime molds, in *Specificity of Embryological Interactions,* Garrod, D. R., Ed., Chapman & Hall, London, 1978, chap. 7.
21. **Barondes, S. H.,** Lectins: their multiple endogenous cellular functions, *Annu. Rev. Biochem.,* 50, 207, 1981.
22. **Rosen, S. D., Reitherman, R. W., and Barondes, S. H.,** Distinct lectin activities from six species of cellular slime molds, *Exp. Cell Res.,* 95, 159, 1975.
23. **Frazier, W. A., Rosen, S. D., Reitherman, R. W., and Barondes, S. H.,** Purification and comparison of two developmentally regulated lectins from *Dictyostelium discoideum.* DISCOIDIN I and DISCOIDIN II, *J. Biol. Chem.,* 250, 7714, 1975.
24. **Simpson, D. L., Rosen, S. D., and Barondes, S. H.,** Pallidin. Purification and characterization of carbohydrate-binding protein from *Polysphondylium pallidum,* implicated in intercellular adhesion, *Biochim. Biophys. Acta,* 412, 109, 1975.
25. **Chang, C.-M., Reitherman, R. W., Rosen, S. D., and Barondes, S. H.,** Cell surface location of discoidin, a developmentally regulated carbohydrate-binding protein from *Dictyostelium discoideum, Exp. Cell Res.,* 95, 136, 1975.
26. **Chang, C.-M., Rosen, S. D., and Barondes, S. H.,** Cell surface location of an endogenous lectin and its receptor in *Polysphondylium pallidum, Exp. Cell Res.,* 104, 101, 1977.
27. **Barondes, S. H.,** Soluble lectins: a new class of extracellular proteins, *Science,* 223, 1259, 1984.
28. **Cooper, D. N. W. and Barondes, S. H.,** Localization of discoidin-binding ligands with discoidin in developing *Dictyostelium discoideum, Dev. Biol.,* 105, 59, 1984.
29. **Bhargava, P. M.,** Regulation of cell division and malignant transformation: a new model for control by uptake of nutrients, *J. Theor. Biol.,* 68, 101, 1977.
30. **Margolskee, J. P., Froshauer, S., Skrinska, R., and Lodish, H. F.,** The effects of cell density and starvation on early developmental events in *Dictyostelium discoideum, Dev. Biol.,* 74, 409, 1980.
31. **Konijn, T. M., van de Meene, J. G. C., and Bonner, J. T.,** The acrasin activity of adenosine-3′,5′-cyclic phosphate, *Proc. Natl. Acad. Sci. U.S.A.,* 58, 1152, 1967.

32. **Wurster, B., Schubier, K., Wick, U., and Gerisch, G.,** Cyclic GMP in *Dictyostelium discoideum.* Oscillations and pulses in response to folic acid and cyclic AMP signals, *FEBS Lett.,* 76, 141, 1977.

33. **Shaffer, B. M.,** The cells founding aggregation centres in the slime mold *Polysphondylium violaceum, J. Exp. Biol.,* 38, 833, 1961.

34. **Gerisch, G.,** Cell aggregation and differentiation in *Dictyostelium, Curr. Top. Dev. Biol.,* 3, 157, 1968.

35. **Sinha, U. K. and Ashworth, J. M.,** Evidence for the existence of elements of a para-sexual cycle in the cellular slime mold, *Dictyostelium discoideum, Proc. R. Soc. London,* B173, 531, 1969.

36. **Newell, P. C. and Mullens, I. A.,** Cell-surface cAMP receptors in *Dictyostelium, Symp. Soc. Exp. Biol.,* 32, 161, 1978.

37. **Gerisch, G., Malchow, D., Roos, W., and Wick, U.,** Oscillations of cyclic nucleotide concentrations in relation to the excitability of *Dictyostelium* cells, *J. Exp. Biol.,* 81, 33, 1979.

38. **Galvin, N. J., Stockhausen, D., Meyers-Hutchins, B. L., and Frazier, W. A.,** Association of cyclic AMP chemotaxis receptor with the detergent-insoluble cytoskeleton of *Dictyostelium discoideum, J. Cell Biol.,* 98, 584, 1984.

39. **Yumura, S., Mori, H., and Fukui, Y.,** Localization of actin and myosin for the study of amoeboid movement in *Dictyostelium* using improved immunofluorescence, *J. Cell Biol.,* 99, 894, 1984.

40. **Garrod, D. R.,** Cell sorting out during differentiation of mixtures of metabolically distinct populations of *Dictyostelium discoideum, J. Embryol. Exp. Morphol.,* 29, 647, 1973.

41. **Steinberg, M. S.,** Does differential adhesion govern self-assembly process in histogenesis? Equilibrium configurations and the emergence of a hierarchy among populations of embryonic cells, *J. Exp. Zool.,* 173, 395, 1970.

42. **Bonner, J. T.,** Evidence for the sorting out of cells in the development of the cellular slime moulds, *Proc. Natl. Acad. Sci. U.S.A.,* 45, 379, 1959.

43. **Forman, D. and Garrod, D. R.,** Pattern formation in *Dictyostelium discoideum,* I. Development of prespore cells and its relationship to the pattern of the fruiting body, *J. Embryol. Exp. Morphol.,* 40, 215, 1977.

44. **Forman, D. and Garrod, D. R.,** Pattern formation in *Dictyostelium discoideum.* II. Differentiation and pattern formation in non-polar aggregates, *J. Embryol. Exp. Morphol.,* 40, 229, 1977.

45. **Smith, E. and Williams, K. L.,** Evidence for tip control of the ''slug/fruit'' switch in slugs of *Dictyostelium discoideum, J. Embryol. Exp. Morphol.,* 57, 233, 1980.

46. **Newell, P. C.,** The development of the cellular slime mould *Dictyostelium discoideum:* a model system for the study of cellular differentiation, in *Essays in Biochemistry,* Vol. 7, Campbell, P. M. and Dickens, F., Eds., Academic Press, London, 1971, 87.

47. **Newell, P. C., Longlands, M., and Sussman, M.,** Control of enzyme synthesis by cellular interaction during development of the cellular slime mold *Dictyostelium discoideum, J. Molec. Biol.,* 58, 541, 1971.

48. **Wilcox, D. K. and Sussman, M.,** Defective cell cohesivity expressed late in the development of *Dictyostelium discoideum* mutant, *Dev. Biol.,* 82, 102, 1981.

49. **Rutherford, C. L., Taylor, R. D., Merkle, R. K., and Frame, L. T.,** Cellular pattern formation: *Dictyostelium discoideum* as a system for a biochemical approach, *Trends Biochem. Sci.,* 7, 108, 1982.

50. **Zucker, R. C. and Lodish, H. F.,** Repetitive DNA sequences cotranscribed with developmentally regulated *Dictyostelium discoideum* mRNAs, *Proc. Natl. Acad. Sci. U.S.A.,* 78, 5386, 1981.

51. **Mangiarotti, G., Ceccarelli, A., and Lodish, H. F.,** Cyclic AMP stabilizes a class of developmentally regulated *Dictyostelium discoideum* mRNAs, *Nature (London),* 301, 616, 1983.

52. **Wilkinson, D. G., Wilson, I., and Hames, B. D.,** Spore coat protein synthesis during development of *Dictyostelium discoideum* requires a low molecular weight inducer and continued multicellularity, *Dev. Biol.,* 107, 38, 1985.

53. **Sternfeld, J. and David, C. N.,** Fate and regulation of anterior-like cells in *Dictyostelium* slugs, *Dev. Biol.,* 93, 111, 1982.

54. **Abe, K., Orii, H., Okada, Y., Saga, Y., and Yanagisawa, K.,** A novel cyclic AMP metabolism exhibited by giant cells and its possible role in the sexual development of *Dictyostelium discoideum, Dev. Biol.,* 104, 477, 1984.

55. **Schaap, P., Konijn, T. M., and van Haastert, P. J. M.,** cAMP pulses coordinate morphogenetic movement during fruiting body formation of *Dictyostelium discoideum, Proc. Natl. Acad. Sci. U.S.A.,* 81, 2122, 1984.

56. **Schaller, K. L., Leichtling, B. H., Majerfeld, I. H., Woffendin, C., Spitz, E., Kakinuma, S., and Rickenberg, H. V.,** Differential cellular distribution of cAMP-dependent protein kinase during development of *Dictyostelium discoideum, Proc. Natl. Acad. Sci. U.S.A.,* 81, 2127, 1984.

57. **Kay, R. R. and Jermyn, K. A.,** A possible morphogen controlling differentiation in *Dictyostelium, Nature (London),* 303, 242, 1983.

58. **Gross, J. D., Bradbury, J., Kay, R. R., and Peacey, M. J.,** Intracellular pH and the control of cell differentiation in *Dictyostelium discoideum, Nature (London),* 303, 244, 1983.

59. **Spemann, H. and Mangold, H.,** Über Induktion von Embryonalanlagen durch Implantation artfremder Organisatoren, *Wilhelm Roux Arch.,* 100, 599, 1924.

INDEX